Hans K. Uhthoff

The Embryology of the Human Locomotor System

With 217 Figures

Springer-Verlag
Berlin Heidelberg New York London
Paris Tokyo Hong Kong Barcelona

Professor HANS K. UHTHOFF MD, FRCS(C)
Chairman and Professor,
Ottawa General Hospital,
Division of Orthopaedic Surgery,
501 Smyth,
Ottawa, Ontario,
Canada K1H 8L6

ISBN 3-540-52028-7 Springer-Verlag Berlin Heidelberg New York
ISBN 0-387-52028-7 Springer-Verlag New York Berlin Heidelberg

Library of Congress Cataloging-in-Publication Data
Uhthoff, Hans K., 1925–. The embryology of the human locomotor system / Hans K.
Uhthoff. p. cm. Essays by Hans Uhthoff in collaboration with the staff, fellows, and
residents of the Division of Orthopaedic Surgery, University of Ottawa. Includes biblio-
graphical references.
ISBN 0-387-52028-7 (U.S. : alk. paper)
1. Musculoskeletal system–Differentiation–Atlases. I. Title. QM602.U37 1990
612.6'4017– dc20

© Springer-Verlag Berlin Heidelberg 1990
Printed in Germany

The use of general descriptive names, registered names, trademarks, etc. in this publication does not imply, even in the absence of a specific statement, that such names are
exempt from the relevant protective laws and regulations and therefore free for general
use.

Product Liability: The publisher can give no guarantee for information about drug dos-
age and application thereof contained in this book. In every individual case the respective user must check its accuracy by consulting other pharmaceutical literature.

Typesetting, printing, and bookbinding: Graphischer Betrieb Konrad Triltsch GmbH,
Würzburg
2124/3145-543210   Printed on acid-free paper

Preface

In this Atlas I want to share with my fellow clinicians the fascination I experienced while discovering the marvels of embryonic development. Why haven't these marvels excited me before? I believe that the use of schemata and drawings or photographs of animal embryos, commonly used in textbooks of embryology, simply did not appeal to me as a clinician. Only actual photographs of human embryos can establish the bond necessary for interaction.

Just imagine the excitement when you find out how many structures you can recognize in a 5-week-old embryo, barely measuring 1 cm in length. But our fascination does not stop here. The progression of changes taking place during the next 3 weeks is so rapid that at the time when the embryo measures 3 cm, all structures familiar to us are not only easily recognizable, but also already in their anatomical position. How can we hide our amazement when we realize that such a state of perfection is present in an embryo a bit longer than the distal phalanx of our little finger?

At 8 weeks the embryonic period ends and the fetal one starts. Although the shape and the relative size of bones, joints, muscles, nerves, and vessels will undergo changes, the basic elements are all in place. This implies that major malformations must develop during the embryonic period.

The Atlas represents a collaborative effort of staff, fellows, and residents of the Division of Orthopaedic Surgery, University of Ottawa. During their studies they became as much fascinated by their observations as I had been. Dr. Caughell of the Division of Plastic Surgery and Dr. Martin of the Department of Anatomy, University of Western Ontario in London contributed to the chapter on the wrist and hand. To all my collaborators I would like to extend my sincerest thanks.

The primordia of muscles and bones cannot be detected before 4 weeks. No illustrations of earlier stages of development are included in this atlas which contains microphotographs chosen from a personal collection of well over 350 spontaneously aborted human embryos and fetuses. This collection was started at the Hôpital Général de Verdun in Montreal and continued at the Ottawa General Hospital. Embryos smaller than 30 mm were serially sectioned in toto, both the sagittal and frontal planes. Although special efforts were made to place the embryos properly prior to sectioning, we did not always succeed and some cuts show an oblique orientation. In larger embryos and in fetuses specific anatomic regions were dissected and embedded separately, often after decalcification in EDTA. Here a transverse plane of sectioning was sometimes added to the other two planes.

The lack of very young embryos in our collection necessitated a trip to the California Primate Research Center in Davis, California where Professor R. O'Rahilly graciously put the Carnegie Collection at our disposal. My only regret is that while in Davis, I could not avail myself of the services of a professional photographer, but had to rely on my amateurish skills. Not surprisingly the quality of these pictures is not comparable to that of the rest of the microphotographs taken so masterly by Stanley Klosevych of the Department of Medical Communication, University of Ottawa. I acknowledge his collaboration with sincere thanks.

I would like to recognize and to thank with deep gratitude Mrs. Wilhelmina Kellam for her continued interest and financial support of our research endeavors. The Atlas became a reality thanks to the constant encouragement and valuable advice given during countless hours by my predecessor as Chief of Orthopaedics, University of Ottawa and close friend, Dr. Jacques Robichon to whom this atlas is dedicated. The final version of the chapters has been preceded by innumerable drafts, all typed and retyped with admirable patience and professional efficiency by Christiane Ménard, my long-standing secretary.

May 1990   H. K. UHTHOFF

Contents

Contributors

TIMOTHY CAREY, MD, FRCS(C),
Fellow,
Children's Hospital of Eastern Ontario,
University of Ottawa,
Ottawa, Ontario, Canada

KEITH A. CAUGHELL, MD, FRCS(C),
Fellow,
Division of Plastic Surgery,
University of Western Ontario,
London, Ontario, Canada

HANI H. EL-KOMMOS, MD, FRCS(C),
Resident,
Division of Orthopaedic Surgery,
University of Ottawa,
Ottawa, Ontario, Canada

MAUREEN A. FINNEGAN, MD, FRCS(C),
Assistant Professor,
Division of Orthopaedic Surgery,
Ottawa General Hospital,
University of Ottawa,
Ottawa, Ontario, Canada

REGGIE CHERINE HAMDY, MD,
Resident,
Division of Orthopaedic Surgery,
University of Ottawa,
Ottawa, Ontario, Canada

TADASHI KAWASHIMA, MD,
Visiting Scientist,
Bone and Joint Research Lab,
University of Ottawa, Canada;
Lecturer,
Department of Orthopaedic Surgery,
University of Niigata,
Japan

JOACHIM F. LÖHR, MD, FRCS(C),
Assistant Professor,
Division of Orthopaedic Surgery,
Ottawa General Hospital,
Ottawa, Ontario, Canada

ALEXANDER H. MARTIN, PhD,
Professor,
Department of Anatomy,
University of Western Ontario,
London, Ontario, Canada

JAMES P. MCAULEY, MD, FRCS(C),
Assistant Professor,
Division of Orthopaedic Surgery,
Ottawa General Hospital,
University of Ottawa,
Ottawa, Ontario, Canada

JACQUES J. ROBICHON, MD, FRCS(C),
Emeritus Professor of Surgery (Orthopaedics),
Division of Orthopaedic Surgery,
University of Ottawa,
Ottawa, Ontario, Canada

HANS K. UHTHOFF, MD, FRCS(C),
Professor and Head,
Division of Orthopaedic Surgery,
University of Ottawa, at the Ottawa General Hospital,
Ottawa, Ontario, Canada

A Guide to Terminology and Organization

Age determination of embryos has always been associated with problems. Although all embryos are routinely measured and weighed, neither length nor weight permit an accurate estimation of age. To remedy this shortcoming, many embryologists have attempted to develop more reliable means. Mall from the Carnegie Institute started a staging system based on structural alterations occurring during embryonic development. Streeter [1] improved this system by dividing the embryonic period into 23 stages. Each stage is characterized by clearly defined and observable details of either the external form or the development of structures. Throughout the Atlas we have attempted to use Streeter's stages as strictly as possible. Given the fact that our Atlas is concerned mostly with the later part of the embryonic period, we limit the short description to stages 15–23.

Stage 15 Nerve trunks stream into the arm bud.
Stage 16 Muscle condensation is evident.
Stage 17 Muscle groups can be identified.
Stage 18 Individual muscles can be seen. Precartilage appears in the humerus.
Stage 19 Straightening of the trunk. Various parts of upper and lower limbs can be identified. Maturation of chondrocytes of humeral anlage.
Stage 20 Hands are approaching lateral margins of nose. No distinction can be made between the cambium and fibrous layer of the periosteum.
Stage 21 Hands flexed at wrist. Feet seem closer to one another. Cambium layer can be recognized.
Stage 22 Fingers of one hand overlap those of the other. Early periosteal sleeve of humerus.
Stage 23 Onset of marrow formation in the humerus.

Cells of the cartilaginous anlages undergo successive changes during their development. Streeter suggested an (arbitrary) division into five phases of cellular differentiation. Phase 1 is characterized by the presence of precartilaginous cells and of scanty intercellular substance. During phase 2 this intercellular matrix gets more abundant and stains deeply. Cells are arranged in rows in a direction perpendicular to the long axis of the future bone. During phase 3, a further increase in the amount of intercellular material can be noted. Chondrocytes become cuboidal and the cytoplasm contains vacuoles. During phase 4, the chondrocytes reach their maximum size. The state of vacuolization is extreme. This phase is often referred to as the phase of hypertrophy. Finally, phase 5 is characterized by an advanced cellular disintegration and by a loss of staining property of the matrix. It is known as the phase of degeneration.

Throughout the Atlas we have aimed at a certain uniformity between the chapters. All chapters describing anatomic regions contain microphotographs of embryos at 6, 7, and 8 weeks and of fetuses at 10, 12, 14, and 16 weeks, and sometimes of 18 and 20 weeks. Microphotographs of full-term fetuses were only exceptionally used. This organization will permit the reader to compare the state of development of various areas.

Finally, we should never forget that the position of limbs in utero, and consequently the position of the joints, is often different from that we are used to see in our daily work. A good example is the hip which is always in a position of flexion. Here the upper part of the femoral head faces posteriorly or posteroinferiorly and therefore is never in contact with the superior dome of the acetabulum.

The final chapter, describing some anatomic variations and malformations, has been added to heighten the clinical interest of the Atlas. It is of note that malformations were found before ossification took place; therefore they cannot be attributed to a primary defect of ossification. Proper ossification cannot take place because the cartilage anlage has not been formed properly.

Although the size of the embryos and fetuses is always stated, as is the magnification of the microphotographs, the reader may experience difficulties in visualizing the actual size of the embryo or fetus. We have therefore added a few representative photographs of specimens ranging in age between 5 and 15½ weeks. Each photograph includes a scale of the 5-week-old embryo for proper comparison.

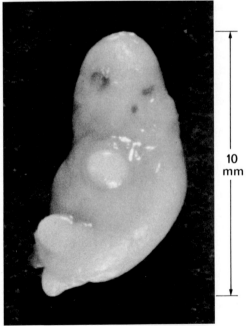

10 mm

Fig. 1. Embryo, 10 mm, 5 weeks, stage 16. Upper limb buds appear as rounded projections. The lower limb buds are less advanced

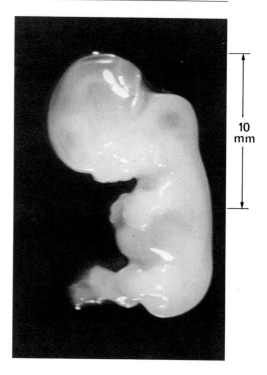

Fig. 2. Embryo, 17 mm, 7 weeks, stage 19. The hand plate with early notching is well seen. The digital plate of the lower limb looks like a fin and has no indentations. The external rotation of the lower limb bud is obvious. The trunk is straightening

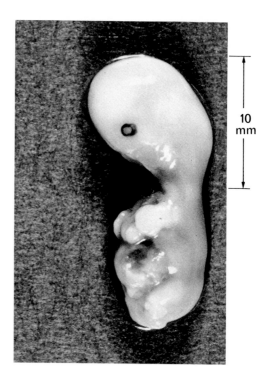

Fig. 3. Embryo, 21 mm, 7 weeks, stage 20. Although this embryo is bigger than that depicted in Fig. 2, it is less developed. The hand is still club-like

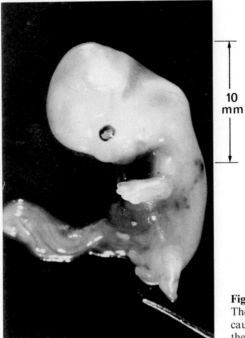

10 mm

Fig. 4. Embryo, 22 mm, 7½ weeks, stage 20. The palms of the hands are facing more caudally. finger rays can be seen. Note again the disproportion between head and body

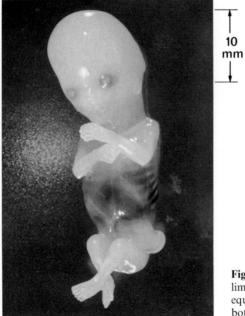

10 mm

Fig. 5. Fetus, 42 mm, 9 weeks. The upper limb is well formed. The foot is still in an equinus position. Note the flexed position of both hips

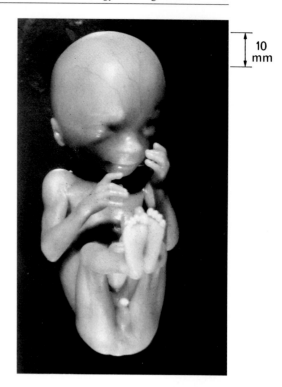

10
mm

Fig. 6. Fetus, 96 mm, 13 weeks. Fingernails
can be seen. The plantar surfaces of both
feet are straight and no equinus persists

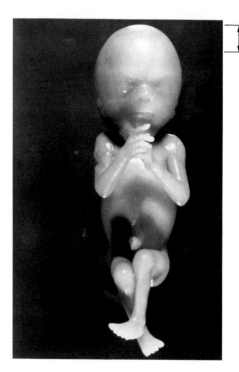

10
mm

Fig. 7. Fetus, 124 mm, 15½ weeks. In these
feet the adduction of both forefeet persists
as well as a slight equinus and an inversion.
Adduction, equinus, and inversion will normally disappear by the end of the 11th week

Reference

1. Streeter GI (1951) Developmental horizons in human embryos. Carnegie Institution of Washington, Washington

Chapter 1

The Development of Limb Buds *

H.K. Uhthoff

Upper Limb

The development of the upper limb bud occurs earlier than that of the lower one.
The first appearance of upper limb buds is noted at stage 12 when the embryo is
approximately 4 weeks of age and measures only 3.5 mm in length. Proliferating
mesoblasts represent the primordium of the upper limb bud which is situated
opposite the 5th–7th cervical somite.

At stage 13 a blastemic condensation in the middle of the limb bud appears.
The ectodermal layer is rather thick (Fig. 1.1). The next important step is growth
of the four lower cervical and the first thoracic nerves streaming into the bud
which at stage 15 remains opposite C5-T1 (Fig. 1.3). Blood circulation in the bud
is established at that stage which corresponds to an age of 33 days. A thin-walled
marginal blood vessel is regularly seen at the tip of the limb bud subjacent to the
apical ectodermal ridge (AER) (Fig. 1.3). This ridge, also known as the placode,
is easily recognizable by the thickening of the ectoderm. The AER of the upper
limb bud is seen first at stage 14 and disappears after stage 17. AERs of upper and
lower buds are of utmost importance in the development of the various elements
of the limbs. The blastemic condensation of the limb buds is referred to as the
skeletomuscular condensation and both structures cannot be differentiated at
stage 15.

During stage 16 muscle and skeletal condensations can be distinguished. The
presence of nerve trunks which obviously invade only muscle condensations
makes easy the recognition of this tissue. During this stage, radial and median
nerves reach the level of the elbow. Also at 5 weeks the humeral anlage shows
evidence of chondrification. In this respect it is important to remember that the
cartilage anlagen of different bones are formed separately.

At 41 days when stage 17 is reached, radius and ulna chondrify (Figs. 1.6 and
1.7). Nerves are also advancing and radial, median, and ulnar nerves reach the
hand plate. At stage 18 the scapula is still at the level of C4-T1 (Fig. 1.9). Carpal
bones chondrify and notching of the digital rays starts.

---

* All but two of the illustrations of this chapter are from the Carnegie Collection

Lower Limb

The lower limb bud starts to develop at stage 13 when the upper limb bud is already seen as a definite, rounded, and projecting appendage. The density of the mesoblastic cells is uniform. A blastemic condensation starts at stage 14 (Fig. 1.2). At stage 15 ingrowth of nerve trunks can be seen. This bud is situated at the level of L1-S1 (Fig. 1.4).

At stage 16 the advancing edge of the bud shows an AER (Fig. 1.5). The AER develops during stage 15 and lasts until stage 18. At stage 16 the lumbosacral plexus is formed by nerve roots from L1–5 and S1–2 and gives rise to major nerve trunks of the lower limb. At stage 17 an early cartilage formation of the femur is found. The lower limb bud is often compared to a fin: its proximal part is more rounded whereas a tapering of the distal half is seen. The latter will form the foot (Fig. 1.8).

At stage 18 the lower limb is still opposite the lower lumbar spine (Fig. 1.11). The cartilage anlagen of the femur and tibia are seen. Again the relatively large size of nerve trunks is surprising, particularly to surgeons used to the anatomy of the child or the adult (Fig. 1.10).

With stage 18, we have terminated the description of the development of the limb buds. It coincides with the appearance of the interzone which heralds the beginning of joint formation and which of course is described in Chap. 3. Thus, at the age of little more than 6 weeks when the embryo only measures 14 mm, even the untrained eye can recognize familiar structures.

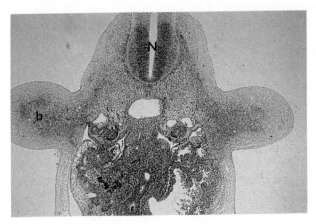

Fig. 1.1. Embryo 8372, 5.6 mm, stage 13. Transverse section. Carnegie Collection. Age ∼4 weeks. The beginning of blastemic condensation (*b*) in the middle of both upper limb buds is obvious. The neural tube shows well (*N*). HPS, ×16

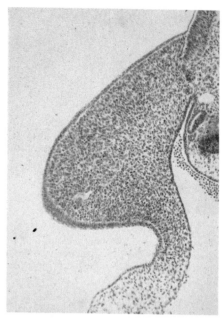

Fig. 1.2. Embryo 6503, 6.3 mm, stage 14. Frontal section. Carnegie Collection. Age ~30 days. The lower limb bud starts to develop. It contains mesoblastic cells which will differentiate into blastemic tissue. H&E, ×40

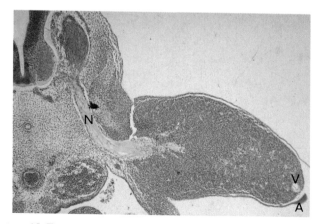

Fig. 1.3. Embryo 721, 9 mm, stage 15. Transverse section. Carnegie Collection. Age ~33 days. The most impressive feature is the ingrowth of nerves (*N*) into the upper limb bud which is still situated at the level of C5-T1. Note the ectodermal thickening at the tip of the limb bud representing the apical ectodermal ridge (*A*); subjacent to it the vessel (*V*). H&E, ×16

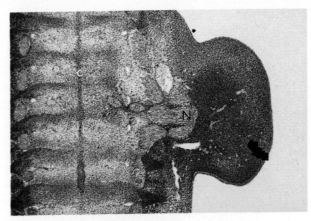

Fig. 1.4. Embryo 6506, 7.5 mm, stage 15. Frontal section. Carnegie Collection. Age ~31 days. The section of the lower limb bud not only shows its position at the level of L1-S1, but also the lateral location. The frontal section cuts through the vertebral bodies and disks in their midpart where the notochord (c) is well seen. Note the nerves (N) entering the bud. Alum cochineal, ×16

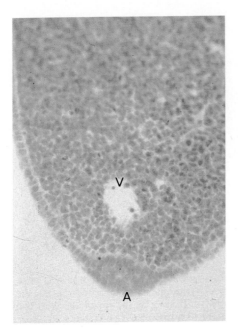

Fig. 1.5. Embryo 6512, 7 mm, stage 16. Transverse section. Carnegie Collection. Age ~37 days. Tip of lower limb bud showing the ectodermal thickening, consisting of several layers of epithelial cells (A). This is often referred to as the apical ectodermal ridge (AER) or placode. Underneath this leading edge of the limb bud we always see a vessel (V). Alum cochineal, ×40

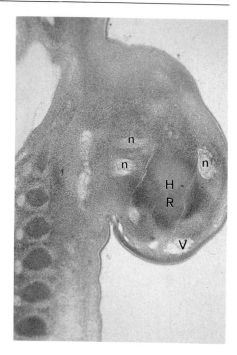

Fig. 1.6. Embryo 485, 13 mm, stage 17.
Frontal section. Carnegie Collection. Age
~6 weeks. The upper limb bud is still in the
lateral position. Chondrification of humerus
(*H*) and radius (*R*) has started. Note also
the presence of three distinct nerves (*n*) and
of blood vessels (*V*). Alum cochineal, ×16

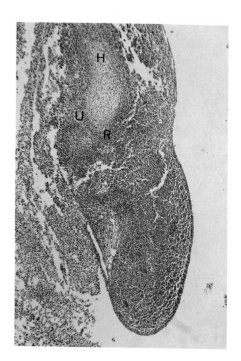

Fig. 1.7. Embryo E15, 102-30, 10 mm, stage
17. Frontal section. Age ~6 weeks. This
section shows well the chondrification of
humerus (*H*), ulna (*U*), and radius (*R*). Cells
surrounding the cartilaginous anlage stain
quite differently from those of the blastemic
condensation further distally. HPS, ×16

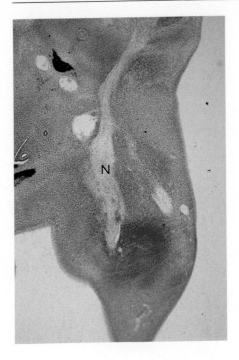

Fig. 1.8. Embryo 4856, 13 mm, stage 17. Frontal section. Carnegie Collection. The lower limb bud is less advanced. Note the huge sciatic nerve (*N*) growing into the bud which is now in a more caudal position. The tapered distal end of the bud will form the foot. India ink is seen in two pelvic vessels, but the vessels of the bud did not fill. Alum cochineal × 16

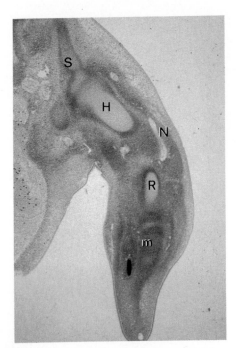

Fig. 1.9. Embryo 6527, 14.4 mm, stage 18. Frontal section. Carnegie Collection. The scapula (*s*) is still at the level of C4-T1. Humerus (*H*) and radius (*R*) are well seen, the metacarpus (*m*) is in an early stage of chondrification. Note the radial nerve (*N*) at the level of the elbow. Alum cochineal × 6.3

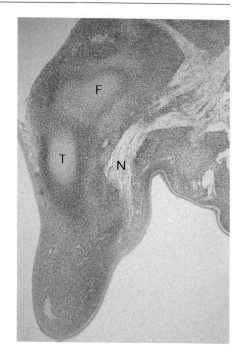

Fig. 1.10. Same embryo. The cartilage anlage of femur (*F*) and tibia (*T*) are well seen.
Note the huge sciatic nerve (*N*). The tapered
end of the bud will become the foot. × 16

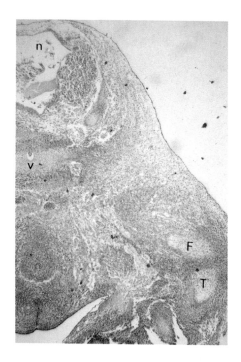

Fig. 1.11. Embryo 117, 6 weeks, 14 mm,
stage 18. Frontal section. The cartilaginous
anlagen of femur (*F*) and tibia (*T*) separated
by a homogeneous interzone are well seen.
Note also a cartilaginous lumbar vertebra
(*v*) including the notochord as well as neural
canal (*n*). Goldner, × 25

References

1. O'Rahilly R, Gardner E (1975) The timing and sequence of events in the development of the limbs in the human embryo. Anat Embryol (Berl) 148:1–23
2. Streeter GI (1951) Developmental horizons in human embryos. Carnegie Institution of Washington, Washington

Chapter 2

The Growth of Tubular Bones

H.K. Uhthoff and J.J. Robichon

Using the femur and its proximal growth plate as an example, we will try to expose as dynamically as possible the changes which take place during bone growth. Basic to this realization is the fact that the entire bone is preformed as a cartilaginous anlage. The first evidence of bone formation is seen at the midpart of the diaphysis where mesenchymal cells lay down a periosteal sleeve of bone (Fig. 2.1). This process is followed by a cellular and vascular invasion of the cartilage model at this level where the chondrocytes have reached the stage of degeneration (Streeter phase 5, Fig. 2.2). This invasion denotes the beginning of enchondral bone formation; in long tubular bones it will move in opposite directions (Fig. 2.3) to reach eventually the metaphyseoepiphyseal region at both ends of the bone where, through a spatial rearrangement of the chondrocytes, the physis or growth plate will be formed. In many short tubular bones only one physis is formed; at the end opposite to the physis, enchondral ossification will progress into the epiphyseal area where, through spatial rearrangement of chondrocytes, a growth plate will be formed under the articular cartilage. It is often referred to as a mini growth plate. The mechanisms of modeling which permit the bone to retain its original shape during growth in length and in width need reemphasizing.

The Two Phases of Enchondral Bone Formation

We propose to distinguish between two different phases of enchondral bone formation which occur in utero during the growth of a long bone. This distinction is mainly based on the difference in the pattern of enchondral bone formation. The first phase of enchondral bone formation starts once a periosteal bony sleeve has formed through direct bone formation. The first phase is characterized by cellular and vascular invasion of the hypertrophic and degenerating chondrocytes in the middle of the anlage at the level of the periosteal sleeve. The second phase starts with the establishment of the actual growth plate at the metaphyseoepiphyseal level. It is characterized by the formation of columns of chondrocytes.

Simultaneously with the enchondral bone formation of the first phase, we can observe the growth of the bony sleeve in the proximal and distal directions until it reaches the level of the metaphyses (Fig. 2.3; Fig. 2.13, number 1). Not only does this sleeve increase in length and thickness, it also undergoes an apposition on the periosteal side and a resorption on the endosteal side, making a widening of the medullary cavity possible. Of course, nature sees to it that the formation of the sleeve always precedes resorption of the degenerating cartilage (enchondral ossification, phase 1); this guarantees a protection of the weak area of hypertrophic cells and thus prevents any mechanical failures (Fig. 2.1).

Enchondral Bone Formation: Phase 1

Let us now look at the enchondral bone formation during the first phase (Fig. 2.13, number 2). Here, hypertrophic and degenerating chondrocytes are replaced by haphazardly arranged primary trabeculae (Fig. 2.3) which in turn will disappear to make room for the hematopoietic marrow. This process progresses toward the extremities of the bone. Although proliferating chondrocytes are present during phase 1, they are not arranged in columns. Consequently, matrix septa lack the longitudinal orientation.

Enchondral Bone Formation: Phase 2

The second phase is heralded by the formation of columns of chondrocytes at the metaphyseoepiphyseal level (Fig. 2.13, number 3). At this stage of development, metaphyseal vessels are present in the region of the hypertrophic and degenerating cells between the septa. Epiphyseal vessels close to the growth plate appear at 20 weeks (Fig. 2.9). Some are crossing the growth plate (Fig. 2.9). We will first give a general outline of the structure and then describe the temporal evolution. Early evidence of columns of chondrocytes, as well as of septae of intercellular matrix, is seen at 10 weeks (Fig. 2.4), but better around 14 weeks (Fig. 2.5); at 20 weeks cartilage cell columns and the septa are well formed (Fig. 2.9). They are even better seen at birth (Fig. 2.11). The thickness of septae increases with time. At the same time the bony cortex will divide into two parts: the diaphyseal bony sleeve and the ring of Lacroix (Figs. 2.7 and 2.8). The interval between the upward moving ring of Lacroix and the diaphyseal sleeve is filled by the peripheral metaphyseal trabeculae of the provisional zone of ossification (Fig. 2.8). Increase in thickness of trabeculae will lead to a transformation of cancellous bone into compact bone (Fig. 2.12). This new cortex is in continuity with the diaphyseal bony sleeve. The more centrally situated trabeculae will be resorbed, thus making room for the medullary cavity. At the metaphysis, which of course is larger in diameter than the diaphysis, the newly formed cortex will be the site of osteoclastic resorption on the periosteal side, and of apposition on the endosteal side, a process well known as modeling. This process observed in the metaphysis is in contradistinction to that seen in the diaphysis where apposition is occurring at the periosteal and resorption at the endosteal side.

The zone of Ranvier, also called the perichondral ossification groove, comprises the bony ring of Lacroix surrounding the physis and its covering loose connective tissue which acts as a cambium layer, assuring growth in width (Fig. 2.5). In turn the cambium layer is covered by a thick layer of fibrous tissue which is continuous with the fibrous layer of the perichondrium and periosteum. It is said that this layer, as well as the ring of Lacroix, contribute to the stability of the physis [1]. The zone of Ranvier extends from the level of the layer of proliferating chondrocytes of the growth plate to the zone of primary trabeculae at the meta-physeodiaphyseal level. At the epiphyseal end of the ring of Lacroix the marginal germinative zone can be seen (Fig. 2.6). This zone produces chondroblasts for diametric growth of the physis, osteoblasts for the ring of Lacroix, and fibroblasts for axial and circumferential growth of the periosteum [2]. The ring of Lacroix occupying the periphery of the physis and the peripherally situated trabeculae overlap, forming the cortex for a short distance, after which the ring undergoes osteoclastic resorption (Fig. 2.7). Evidently, the processes of accretion at the ring of Lacroix at the epiphyseal level and its resorption at the metaphyseodiaphyseal level, assure the movement in space of the ring necessary for growth in length. Osteoclastic resorption at the metaphyseal end of the ring of Lacroix does not always occur. At some sites the ring is continuous with the cortex formed by primary trabeculae (Fig. 2.10).

Between 9 and 10 weeks blood vessels invade the proximal femoral epiphysis at the sites of synovial reflections. At 13 weeks epiphyseal blood vessels have grown deeper into the epiphysis and shortly after they communicate with the metaphysis. These transphyseal vessels will decrease in number at 8 months postpartum and disappear at 18 months [3].

The architecture of the growth plate needs special attention. Our largest specimen obtained at full term still shows a straight growth plate without any indentations or spikes, also called mamillary processes (Fig. 2.11). They appear only when the epiphyseal nucleus develops after birth; they protect the physis against shear stresses.

Enchondral bone formation of the epiphyses of long tubular bones follows a pattern similar to that seen in diaphyses. No periosteal bone formation, however, will precede the vascular and cellular invasion of the central area where chondrocytes hypertrophy and then undergo degeneration (Fig. 2.13, number 4). Once the enchondral bone formation has reached the level of the articular cartilage, the epiphyseal growth plate (Fig. 2.13, number 5) will be established. The timing of the appearance of the epiphyseal ossific nucleus varies from bone to bone. In short tubular bones which possess only one physis, an epiphyseal growth plate will be formed at the end opposite to the physeal growth plate.

Although the first and second phases of enchondral ossification as previously described are different, they are intimately related. Any abnormality of either process can result in well-known pathologic conditions. Examples of diaphyseal growth disturbances are osteogenesis imperfecta and congenital pseudarthrosis whereas scurvy and dwarfism are examples of physeal disturbances.

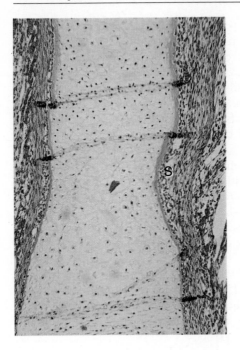

Fig. 2.1. Embryo 49N, 8 weeks, 32 mm. Frontal section. Formation of the periosteal bony sleeve (*S*) and maturation of chondrocytes into Streeter phases 4 and 5 are seen. As yet no evidence of invasion of the cartilage anlage by loose connective vascular tissue. Goldner, × 100

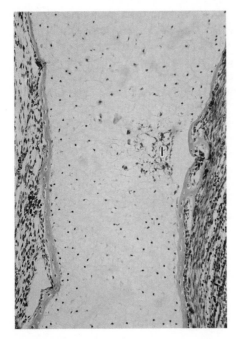

Fig. 2.2. Embryo 50N, 8 weeks, 30 mm. Frontal section. Phase 1 of enchondral bone formation. Cellular invasion (*I*) of femoral diaphysis and resorption of degenerating chondrocytes, in part by multinucleated giant cells. No blood vessels can be definitely identified, but erythrocytes are present. Azan, × 100

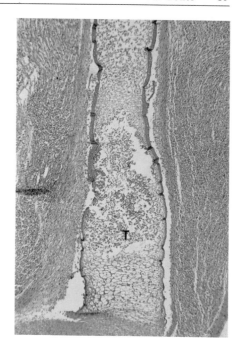

Fig. 2.3. Fetus 51N, 9½ weeks, 45 mm. Sagittal section. a rather haphazard enchondral bone formation progresses distally and proximally. Presence of some primary trabeculae (*T*). The bony periosteal sleeve extends to the zone of proliferating chondrocytes. Goldner, ×16

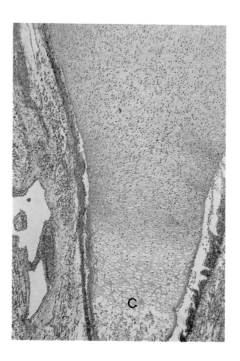

Fig. 2.4. Fetus 112, 10 weeks, 53 mm. Frontal section. Some hypertrophic chondrocytes (*C*) are lining up in columns. Goldner, ×16

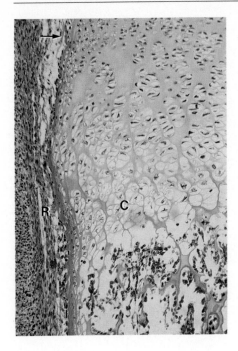

Fig. 2.5. Fetus 108N, 13½ weeks, 105 mm. Frontal section. Early columnar arrangement of chondrocytes (*C*) with scanty intercellular matrix. *R* indicates ring of Lacroix; *arrow* points to site of accretion of the bony ring. Goldner, × 100

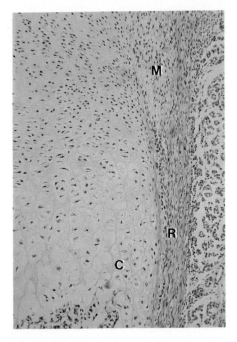

Fig. 2.6. Fetus 92, 16 weeks, 130 mm. Frontal section. Phase 2 of enchondral bone formation. The cell columns (*C*) are now evident, but the intercellular septae are still small. The ring of Lacroix (*R*) is well developed; *arrow* points to the area of accretion. Osteoblasts are seen in the loose-celled zone of Ranvier, assuring the growth in width. The marginal germinative zone (*M*) is well seen. HPS, × 100

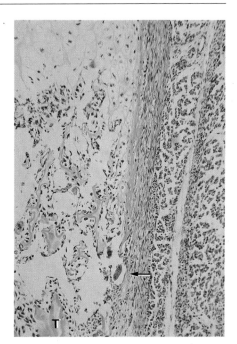

Fig. 2.7. Same fetus as in Fig. 5. This microphotograph taken further distally shows details of the lower end of the ring of Lacroix and of metaphyseal modeling. Osteoclasts (*arrow*) remove the bony ring of Lacroix. Slightly distal and medial to it primary trabeculae (*T*) are seen forming the new cortex which undergoes osteoclastic resorption on its outer aspect and bone apposition on the inner face, both processes contributing to metaphyseal modeling. × 100

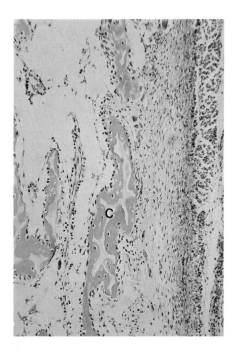

Fig. 2.8. Same fetus as in Fig. 5. This microphotograph has been taken further distally and shows better the process of metaphyseal modeling. Through this process a cortex will be formed which still contains the cartilaginous cores (*C*) of primary trabeculae. × 100

Fig. 2.9. Fetus 115N, 20 weeks, 185 mm. Frontal section. The matrix septa are now thicker. The white islands within the trabeculae represent the persisting cartilaginous matrix (cartilage cores). The entire extent of the ring of Lacroix is well shown; *arrows* point to its beginning and its end. Note the multitude of epiphyseal vessels; one cartilage canal is crossing the physis (*V*). Goldner, ×40

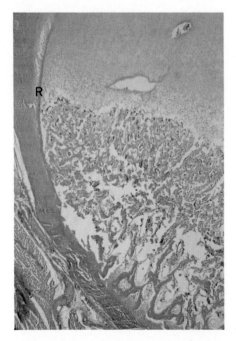

Fig. 2.10. Fetus 174N, 21 weeks, 200 mm. Frontal section. This shows well the lateral aspect of the growth plate and the overlying epiphysis. The ring of Lacroix (*R*) is well formed. Note the presence of hematopoietic cells between the trabeculae. No mamillary processes are present. Note the metaphyseal modeling. Goldner, ×40

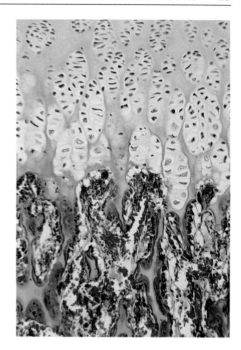

Fig. 2.11. Fetus A1556, full term, 320 mm, Frontal section. The columnar arrangement of the chondrocytes of the physis is obvious as well as the formation of primary trabeculae. The growth plate is straight. Goldner, × 100

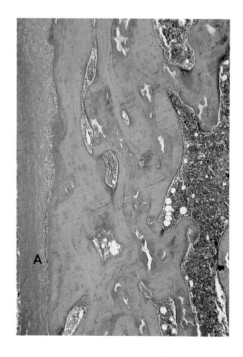

Fig. 2.12. Fetus A1556, full term, 320 mm. Frontal section. This microphotograph illustrates the formation of compact bone. New bone fills the spaces between primary trabeculae. Periosteal apposition is visible further distally (*A*). Some cartilage cores can still be seen. Goldner, × 40

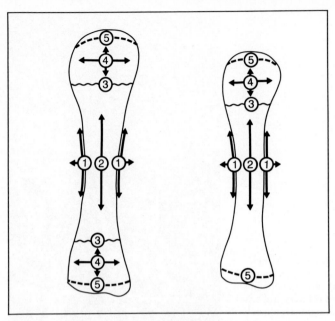

Fig. 2.13. Schematic outline of the progressive stages of enchondral ossification of a long and a short tubular bone. Periosteal diaphyseal bony sleeve (1); enchondral bone formation, phase 1 (2); enchondral bone formation, phase 2 (the physeal growth plate) (3); epiphyseal enchondral bone formation (4); epiphyseal growth plate (5)

References

1. Ogden J, Rosenberg LC (1988) Defining the growth plate. In: Uhthoff HK, Wiley JJ (eds) Behavior of the growth plate. Raven, New York, pp 1–15
2. Speer DP (1988) The growth plate: structure and function. In: Dee R, Mango R, Hurst LC (eds) Principles of orthopaedic practice, vol 1. McGraw-Hill, New York, pp 96–109
3. Trueta J (1959) The three types of acute hematogenous osteomyelitis. J Bone Joint Surg [Br] 41:671–680

Chapter 3

The Development of the Articular Cartilage

H.K. UHTHOFF

The successive changes the articular cartilage of a given synovial joint undergoes during intrauterine life are similar for all diarthrodial articulations. It was therefore decided to describe such development in a special chapter to avoid unnecessary repetition. The illustrations correspond to sequential steps of development, using various joints as examples.

It is well established that the successive stages of development occur earlier in the upper limb than in the lower limb. Moreover, the stage of development progresses in a proximal to distal direction. This order, however, does not necessarily apply to the process of cavitation.

The future joint, situated between chondrifying anlagen, appears as a homogeneous blastemic cell condensation with uniform nuclei (Fig. 3.1). This condensation of cells is often referred to as the uniform interzone. During the next phase three layers of the interzone can be distinguished. The upper and lower layers are dense and their cells are round; later, they will be transformed into chondrogenous layers forming eventually the articular cartilage. These layers are in continuity with the perichondrium of the anlagen. The intermediate layer of the interzone which is less dense and not connected with the perichondrium contains cells with elongated nuclei which are oriented parallel to the joint line (Fig. 3.2). This layer will form all intra-articular structures such as menisci and certain ligaments. The transformation of round into elongated nuclei is sometimes retarded (Fig. 3.4).

We would like to divide the formation of the actual joint space into two successive stages. During the first stage, also called the stage of cavitation, cleft-like spaces appear in the intermediate loose-celled layer. In our specimens we could not find a predilection for a peripheral or central beginning (Fig. 3.3). Serial sections of a given joint regularly showed that the process of cavitation is quite different at different sites as seen in sections going through the hip (Fig. 3.5a, b) and through the elbow joint (Fig. 3.6a, b). The number and extent of the cavitations gradually increase; they eventually coalesce. The second stage consists of the formation of the joint cavity. Early on, strands of cells are still seen crossing the joint, and the surface of the articular cartilage remains irregular and very cellular (Fig. 3.7).

In our material, cavitation of the hip joint became evident at stage 18 (age 6½ weeks, 15 mm). Pertinent publications, however, place its appearance at later stages [1, 2]. We observed cavitation of the shoulder joint during stage 23 (8 weeks, 30 mm) whereas Olah and Ludwig [5] saw it in embryos 28–29 mm in length and Gardner and Gray [3] observed its appearance in an embryo of 25 mm. The report of Gray and Gardner [2] in relation to the elbow corresponds to our findings. In fact, we saw cavitation at 7 weeks (stage 19, 20 mm). Cavitations are neither confluent nor extensive enough to permit motion. Cavities, on the other hand, extend far enough to allow motion. This is in keeping with the statement by O'Rahilly [6] "that early cavitation depends on an intrinsic mechanism but that articular motion is required for the full differentiation and maintenance of the cavity." We believe that the process of cavitation is independent of movements whereas a joint cavity can only form in the presence of movement.

At age 11 ½ weeks the cells of the chondrogenous layers of the coxofemoral joint become indistinguishable from deeper lying cells. The lamina splendans can be seen (Fig. 3.8). The joint surface is fairly regular. With further development the superficial chondrocytes take a tangential orientation, the nuclei become slender, whereas deeper cells keep their round aspect and the joint surface is now regular (Fig. 3.9). A concomitant increase in intercellular matrix becomes obvious.

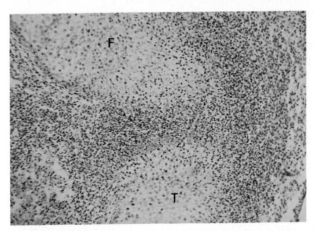

Fig. 3.1. Knee of embryo 117N, 6 weeks, 14 mm, stage 18. Sagittal section. This shows blastemic cells of uniform size, density, and shape between the cartilaginous anlagen of femur (*F*) and tibia (*T*). Goldner, × 100

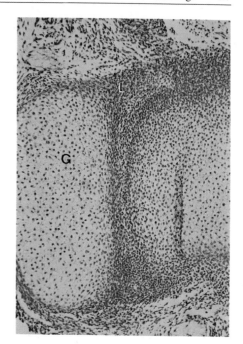

Fig. 3.2. Shoulder of embryo 47N, 6½ weeks, 15 mm, stage 19. Frontal section. The three layers of the interzone start to appear through a loosening of the middle layer. Condensation of cells at both extremities of the glenoid (*G*) indicate an early formation of labra (*L*). Azan, ×100

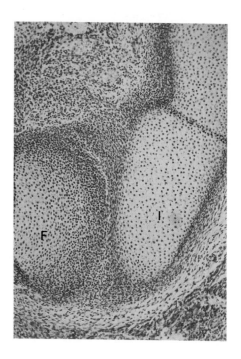

Fig. 3.3. Hip of embryo 47N, 6½ weeks, 15 mm, stage 18. Sagittal section. A decrease in cellular density of the middle layer of the interzone is evident. Cavitation is starting. The perichondrium of the femur is in continuity with the chondrogenous layer of the femoral head (*F*). Ilium (*I*). HPS, ×100

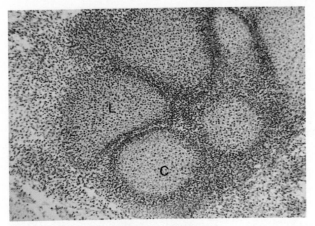

Fig. 3.4. Carpus of embryo 75, 8 weeks, 27 mm, stage 21. Oblique section. Showing the three layers of the lunatocapitate joint. Cells in its intermediate layer are loosely packed, but their nuclei remain round. The interzones of other joints still remain homogeneous. Lunate (*L*); capitate (*C*). Goldner, ×100

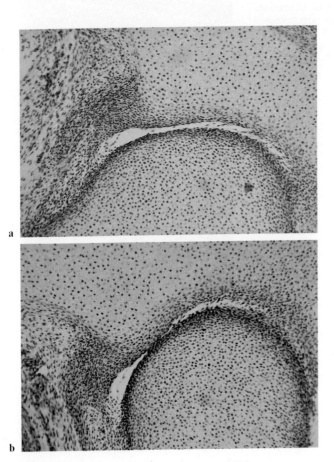

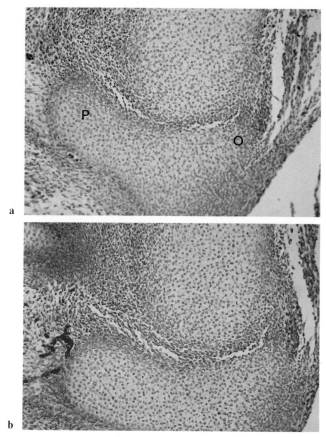

Fig. 3.6 a, b. Elbow of embryo 245N, 7 weeks, 21 mm, stage 19. Sagittal section. **a** The cavitation reaches the olecranon (*O*), but does not extend to the coronoid process (*P*). Goldner, × 100. **b** In this more lateral section the cavitation goes to the coronoid process, but is not complete in the midpart. Strands of cells crossing the joint are particularly well seen. These differences speak against mechanical factors responsible for joint formation. Azan, × 100

Fig. 3.5 a, b. Hip of embryo 119N, 7 weeks, 20 mm, stage 19. Frontal section. Showing (**a**) a more central section and (**b**) a section going through the posterior part of the same joint. The process of cavitation is more advanced in the central part. Some strands of cells remain inside the joint. The articular surface is irregular. Goldner, × 100

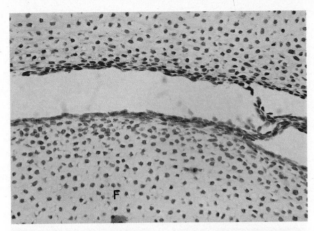

Fig. 3.7. Hip of embryo 193N, 7½ weeks, 25 mm, stage 23. Sagittal section. Some cells of the superficial layer are still rather round. The thickness of the chondrogenous layers has decreased. The joint surface is irregular and strands of cells between both surfaces remain. The chondrogenous layer of the femoral head at the extreme right has been detached accidentally during processing. Femoral head (*F*). Goldner, × 200

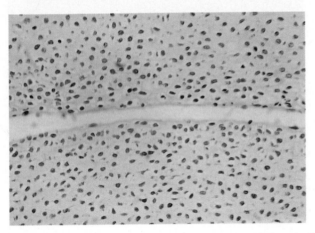

Fig. 3.8. Hip of fetus 52N, 11½ weeks, 70 mm, CR length. Sagittal section. The articular surfaces are smooth. The lamina splendans stains well. Superficial cells are still round. Goldner, × 200

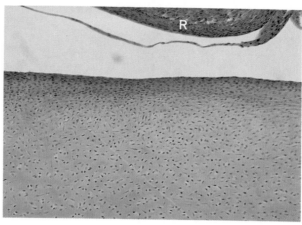

Fig. 3.9. Shoulder joint of fetus 174, 21 weeks, 200 mm. Frontal section. Articular surface of the humerus: note the tangential arrangement of superficial chondrocytes which are now flat. Rotator cuff (*R*). Goldner, × 100

References

1. Fricker HS (1966) Beitrag zur Frühentwicklung des Os coxae beim Menschen. Acta Anat (Basel) 65:522–550
2. Gardner E, Gray DJ (1950) Prenatal development of the human hip joint. Am J Anat 87:163–211
3. Gardner E, Gray DJ (1953) Prenatal development of the human shoulder and acromioclavicular joints. Am J Anat 92:219–276
4. Gray DJ, Gardner E (1951) Prenatal development of the human elbow joint. Am J Anat 88:429–469
5. Olah AJ, Ludwig KS (1971) Die Embryonalentwicklung des Sternoclaviculargelenkes beim Menschen. Acta Anat (Basel) 78:383–405
6. O'Rahilly R (1985) The embryology of joints. In: Verbruggen G, Veys EM (eds) Degenerative joints, vol 2. Elsevier, Amsterdam
7. O'Rahilly R, Gardner E (1978) The embryology of movable joints. In: Sokoloff L (ed) The joints and synovial fluid, vol 1. Academic, New York, pp 49–103

References

Chapter 4

The Early Development of the Spine

H.K. UHTHOFF

During early embryonic life (day 20, stage 9) somites start to develop on each side of the neural tube; they will increase in number during the next 2 weeks. The somites will give rise to the development of skeletal muscle and vertebrae [1]. The part which will form the muscle is called the myotome. The part forming the vertebra is called the sclerotome which will grow towards the midline. Between sclerotomes an intersegmental artery is present. This gives rise to a primitive (first) segmentation which will become obvious at stage 12 when the embryo measures 3.5 mm in length. Each segment has been called a primitive vertebra [3], a structure quite different from the definitive vertebra. The most important landmarks in the distinction between primitive and definitive vertebrae are their respective locations, the position of the artery, and the location of vertebral pedicle and nerve root. Whereas in primitive vertebrae the artery appears to separate one vertebra from the other, in the definite vertebra it lies in the middle of the vertebral body. The sclerotomes lie lateral to the neural tube at stage 12. Therefore, the primitive vertebrae and their segmentation are only evident in paraxial sagittal sections. For this reason we prefer to call the primitive vertebra a primitive segment. The aspect of segmentation is heightened by conspicuous skin ridges (Fig. 4.1). At that stage, the cells of the primitive segment are uniform in shape and density. One stage later, cells of the cranial part of the primitive segment are less densely packed than those of the caudal half. Both halves are separated by an intrasclerotomic fissure, well seen at stage 14 [2]. In frontal sections one can appreciate that the alternation in density is only seen lateral to the midline. At the midline the cell density remains uniform (Fig. 4.2).

At stage 15 when the embryo measures 8 mm in length, this difference becomes more obvious. Although this is evident in frontal sections, it can be better demonstrated in serial sagittal sections. At the level of the ganglia a striking difference in cell density between the cranial and caudal halves of the primitive segment can be seen (Figs. 4.3, 4.4 and 4.12, section 1). As we have stated, the intersegmental artery constitutes the division between primitive segments. Furthermore, it seems that nerve root and ganglion lie behind its superior, that is the loose-celled, half whereas the anlage of the vertebral pedicle lies behind its inferior, that is the dense-celled, half (Fig. 4.12, section 1). However, as Figure 4.3 shows, this arrangement is less clear-cut than many schematic drawings make us believe. It is to be remembered that in the definitive vertebra the pedicle is situated behind the superior half of the vertebra whereas the nerve root lies behind the inferior half

(Fig. 4.10). In serial sagittal sections closer to the midline of the body (Fig. 4.12, section 2), which pass just lateral to the neural tube, alternating loose- and dense-celled zones can still be seen (Fig. 4.5). At cursory inspection, the artery seems to lie between the dense-celled zone cranially and the loose-celled zone caudally. However, at closer inspection the loose-celled zone spreads cranially beyond the artery (Fig. 4.6). Going further medially (Fig. 4.12, section 3), the artery can now be seen in the middle of the loose-celled zone (Figs. 4.7 and 4.8). The next serial sagittal section going exactly through the midline of the body shows the notochord ((Fig. 4.12, section 4; Fig. 4.9). Here, evidence of segmentation and differences in cell density are lacking. Also, at this level segmental arteries cannot be seen anymore. Obviously, the process of segmentation which proceeds from lateral to medial, has not reached the midline yet, but will do so with further growth. At that time the artery will already be situated in the middle of the loose-celled zone.

An understanding of the spatial events occurring during the change from primitive segment to definitive vertebra is essential for the following discussion of Remak's resegmentation theory. Remak [3] studied chick embryos where he made observations based on lateral transillumination of entire chicks. Obviously, this did not permit him to assess the lateral-medial migration. He interpreted the changes as being temporal in nature. This in turn necessitated an explanation of the obvious change in the relationship between artery and surrounding tissue as well as the relative change in location of the future pedicle and nerve root. Simply stated the gradual shift of the loose-celled half in relation to the artery, and thus to the anlage of the pedicle over a certain period of time, was interpreted by him as evidence of resegmentation. It is to be remembered, however, that the site where the definitive vertebra will form is different from the more lateral location of the primitive segment. Moreover, the artery always lies in the middle of the loose-celled zone which is the anlage of the definitive vertebra. Consequently, there is no need for a resegmentation to occur. Others, among them Verbout [6], have voiced this opinion before. It seems obvious from our observations that the changes which Remak [3] had thought to be in time are in reality taking place in space.

With further growth when the embryo reaches stage 17 and measures 12.5 mm, it becomes evident that the loose-celled half of the primitive segment will have progressed to become the definitive vertebra (Fig. 4.10). The pedicle is seen at the cranial part of the vertebra and the nerve root lies posterior to the caudal part. This arrangement is present from the moment when the process of segmentation reaches the site of the future, definitive vertebra.

In summary, two striking features characterize the period of primitive segments. One is the segmentation of the ectoderm lateral to the neural tube, giving rise to the transverse skin ridges. The other feature concerns the aspect of segmentation caused by the alternation between loose-celled and dense-celled zones. A primitive segment consists of a loose-celled half cranially and a dense-celled half caudally. The artery constitutes the border between neighboring segments. The position of the artery changes gradually, as can be seen when viewing sections closer to the midline; there it will be found in the middle of the loose-celled half. These facts did not receive sufficient attention in our earlier studies in which we endorsed Remak's resegmentation theory [5]. Realizing our error we now agree with Verbout [6], as well as with O'Rahilly and Benson [1] that the definitive vertebrae develop from unsegmented anlage material. An easy recognition of all elements of the spine becomes possible at stage 18. Vertebral bodies which start to chondrify, intervertebral disks, segmental arteries, and the ribs can be easily seen (Fig. 4.11).

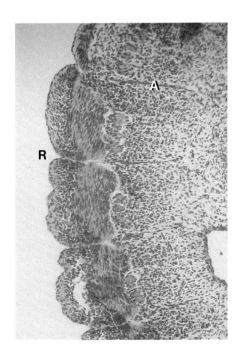

Fig. 4.1. Embryo E13, 7 mm, stage 14. Sagittal section. The sclerotomes giving rise to the primitive segments are easily recognized. They are separated by intersegmental arteries (*A*) and the appearance of segmentation is heightened by conspicuous skin ridges (*R*). The cellular density is rather uniform. H & E, × 120

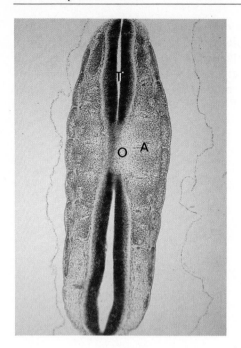

Fig. 4.2. Embryo 1075, 6 mm, stage 13. Frontal section. Carnegie Collection. Sclerotomes lateral to the neural tube (*T*) are well seen and arteries (*A*) define the interval between the sclerotomes. Where the section passes anterior to the neural tube (at the midline) no segmentation (*O*) is as yet evident. H & E, × 16

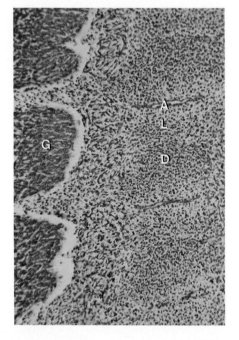

Fig. 4.3. Embryo 55N, 8 mm, stage 15, ~5 weeks (Fig. 4.12, section 1). Figures 4.3–4.9 are from the same embryo and represent serial sagittal sections going from lateral to the midline of the body. This section goes through the level of the ganglia (*G*) which, of course, are situated laterally; they lie behind the loose-celled half (*L*) of the sclerotome. Cranial to the loose-celled half the artery (*A*) can be seen. The caudal dense-celled half (*D*) of the primitive segment is well seen. Cells between the ganglia could well represent the future pedicle. They are in continuity with that part of the loose-celled half which is situated cranially to the artery. Note the size of the ganglia when compared with that of the segments. HPS, × 120

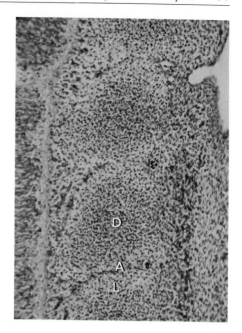

Fig. 4.4. This section, taken approximately at the same level, also shows the artery (*A*) lying between the dense-celled area (*D*) cranially and the loose-celled area (*L*) caudally. Azan, ×120

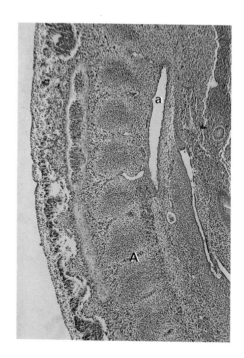

Fig. 4.5. Same embryo as Fig. 4.3 (Fig. 4.12, section 2). This more medially lying section passes just lateral to the neural tube. The arteries (*A*) are clearly seen and they seem to divide the loose-celled area into a small cranial and a bigger caudal part. The dense-celled half of the primitive segment is situated caudal to the loose-celled half. In the upper part of this section, the aorta can be seen (*a*). HPS, ×40

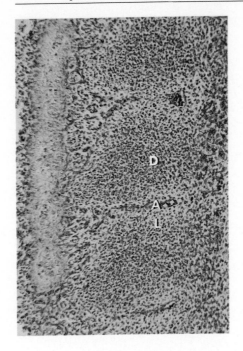

Fig. 4.6. Higher magnification of part of Fig. 4.5, showing the relation of the artery (*A*) to the loose-celled area (*L*). Dense-celled area (*D*). × 120

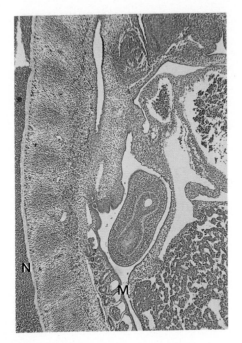

Fig. 4.7. Same embryo as Fig. 4.3 (Fig. 4.12, section 3). This section lies more medial than Figs. 4.5 and 4.6, and passes through the neural tube (*N*) slightly lateral to the midline; it shows the artery in the middle of the loose-celled area. This zone will become the vertebra. Evidently a gradual cranial shift of the loose-celled area in relation to the artery has taken place. Anterior to the caudal part of the spine the mesonephros (*M*) can be seen. HPS, × 40

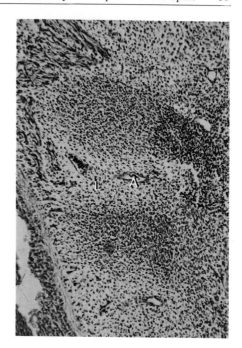

Fig. 4.8. Same embryo as Fig. 4.3. This higher magnification of Fig. 4.7 shows the exact position of the artery (*A*) in the middle of the loose-celled area (*L*). × 120

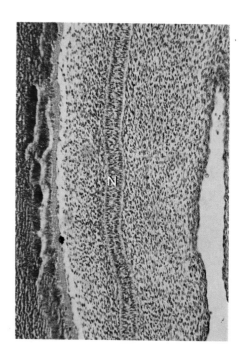

Fig. 4.9. Same embryo as Fig. 4.3 (Fig. 4.12, section 4). This section goes exactly through the midline of the body, as evidenced by the presence of the notochord (*N*). There is no sign of segmentation. Cells are uniform in density. Azan, × 100

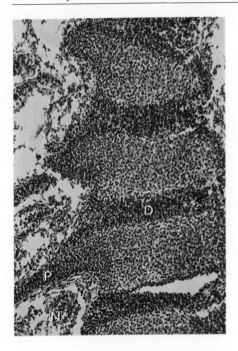

Fig. 4.10. Embryo 82, 12.5 mm, stage 17, ~6 weeks. Sagittal section. Definitive vertebra. The position of the pedicle (*P*), posterior to the cranial part of the vertebral body, is seen. Note the early chondrification of the body. The nerve root (*N*) is seen caudal to the neural arch. The dense-celled area constitutes the intervertebral disk (*D*). Goldner, ×100

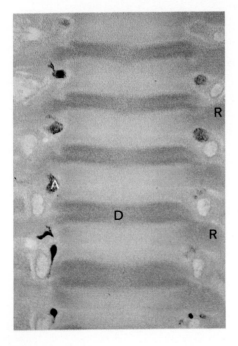

Fig. 4.11. Embryo 492, 16.8 mm, stage 18. Frontal section. Carnegie Collection. This clearly shows that the more lateral dense-celled zones forming the ribs (*R*) are slightly caudal to the dense-celled area of the intervertebral disc (*D*). The segmental arteries (*A*) are prominent owing to an india ink injection. Their position at the midpart of the vertebral body is obvious. Alum cochineal, ×16

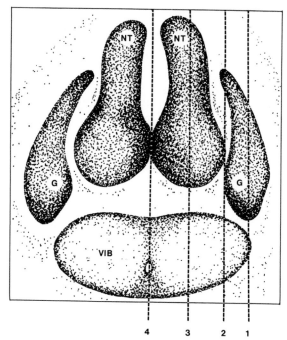

Fig. 4.12. Schema based on a transverse section of embryo 6517 of the Carnegie collection (stage 16) to show the level of histologic sections of Fig. 4.3–4.9. Ganglion (*G*), neural tube (*NT*), vertebral body (*VB*)

References

1. O'Rahilly R, Benson DR (1985) The development of the vertebral column. In: Bradford DS, Hensinger RM (eds) The pediatric spine. Thieme, Stuttgart, pp 3–18
2. O'Rahilly R, Meyer DB (1979) The timing and sequence of events in the development of the human vertebral column during the embryonic period proper. Anat Embryol (Berl) 157:167–176
3. Remak R (1885) Untersuchungen über die Entwickelung der Wirbelthiere. Reimer, Berlin, pp 40–44
4. Sensenig EC (1949) The early development of the human vertebral column. Contrib Embryol Carnegie Inst 33:21–41
5. Tanaka T, Uhthoff HK (1981) Significance of resegmentation in the pathogenesis of vertebral body malformation. Acta Orthop Scand 52:331–338
6. Verbout AJ (1976) A critical review of the "Neugliederung" concept in relation to the development of the vertebral column. Acta Biotheor [A] (Leiden) 25:219–258

Chapter 5

The Development of the Cervical Spine

R.C. HAMDY and H.K. UHTHOFF

The development of the cervical spine can be divided into three stages [4, 7, 8]:

1. Segmentation, occurring between 3 and 6 weeks
2. Chondrification, starting at 6 weeks
3. Ossification, starting at 10–12 weeks

The stage of segmentation during which somites from each side migrate towards the neural tube to form the vertebral column, will not be discussed here as it was dealt with in Chap. 4.

Stage of Chondrification

The stage of chondrification starts at about 6 weeks. During that stage we can observe that vertebral chondrification occurs as one cartilage anlage. This fact is important since some previous descriptions indicated that chondrification of the vertebral body occurs through two separate cartilage centers which later coalesce [1, 4, 8, 9]. The basic shape of the cervical vertebrae is already formed by 7 weeks (Figs. 5.1 and 5.2), following which the cartilaginous vertebrae continue to grow in size.

Stage of Ossification

Ossification starts in both posterior elements around the 10th–11th week (Fig. 5.5). The vertebral bodies start to ossify 2–3 weeks later (Fig. 5.8). The cervical vertebrae (except those of the atlas and the axis) ossify from three primary ossification centers: two for the posterior elements or neural arches and one for the body [1, 3, 4, 8].

Bone formation in the vertebral bodies is preceded as usual by invasion of the cartilage anlage by vascular channels. We found that vascular invasion takes two different forms:

1. Vessels invading the central part of the body where chondrocytes have already undergone degeneration (Streeter phase 5). Enchondral bone formation phase 1 immediately follows vascular ingrowth (Fig. 5.8).

2. Vessels invading the peripheral parts of the vertebral bodies. No degeneration and no bone formation occurs around these vessels. They are termed cartilage canals and serve as a source of nutrition through diffusion (Fig. 5.6).

Atlanto-Axial Complex

The atlas develops from two primary ossification centers; one for each of the two neural arches [4, 8]. The axis develops from five primary ossfication centers [4, 8]:

Two for the neural arches. They appear between 8 and 10 weeks (Figs. 5.5 and 5.8)

One for the body of the axis, appearing around the 12th week

Two for the odontoid, appearing around the 20th week

In some specimens a layer of increased cell density could be seen in the midpart of the odontoid process (Figures 5.1, 5.2, 5.3 and 5.4). Its persistence after birth could be one of the causes of os odontoideum. This layer lies above the level of the superior articular facets of the axis in contrast to the neurocentral synchondrosis which is a cartilaginous band present at birth; it lies between the base of the odontoid process and the body of the axis, below the level of the superior articular facets [2].

Relationship Between the Odontoid Process and the Occiput

As already observed by O'Rahilly et al. in 1983 [5] we also noticed that the odontoid occupies a high position in relation to the basal plate and foramen magnum (seen between 7 and 16 weeks). Persistence of this cranial position of the odontoid process in relation to the occiput may explain some atlanto-axial-occipital abnormalities such as basilar impression.

Blood Supply of the Odontoid

As shown in Figs. 5.7 and 5.9, the apical, alar and transverse ligaments have a very rich vascular network, indicating that (at least in the embryonic and early fetal periods) a major part of the blood supply to the odontoid comes through these ligaments. However, it is believed that these same ligamentous vessels play a minor role in the blood supply of the adult odontoid, as previously shown by Schatzker et al. in 1975 [6].

Ligaments of the Atlanto-Axial Joint Complex

The six ligaments that are attached to the atlanto-axial complex are usually developed by the 8th week:

1. The apical ligament, connecting the tip of the odontoid to the foramen magnum (Figs. 5.1, 5.3, and 5.9)

2. The alar ligaments (vein ligaments), connecting the sides of the odontoid to the foramen magnum (Figs. 5.1, 5.3, 5.4, and 5.9)
3. The transverse ligament in contact with the posterior aspect of the odontoid (Figs. 5.4, 5.7, and 5.9)
4. The inferior and superior parts of the cruciform ligament (Fig. 5.7)
5. The atlanto-occipital membrane (Fig. 5.1)
6. The accessory ligaments, extending from the bases of the lateral masses of the atlas to the base of the dens and the body of the axis

The importance of these ligaments to the blood supply of the odontoid has already been mentioned.

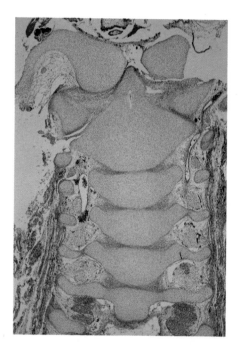

Fig. 5.1. Embryo 150N, 7 weeks, CR 18 mm. Frontal section. The section clearly shows the occiput, both arches of the atlas, the axis as well as the bodies and lateral elements of C3–C6. There is a zone of increased cellular density in the midpart of the odontoid. Part of the notochord is also clearly seen at that level. The atlanto-occipital membrane and the atlanto-axial ligaments are well demonstrated. A single cartilage anlage, including the body and both transverse processes, is clearly seen at the level of C-6. The vertebral artery can also be recognized. Goldner, × 20

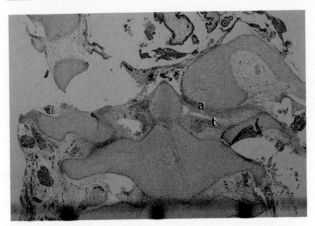

Fig. 5.2. Same embryo as Fig. 5.1. The higher magnification shows better the ligaments and the relation between the occiput, atlas, and axis. Alar ligament (*a*), transverse ligament (*t*). Of note is the proximal position of the tip of the odontoid in relation to the occiput. Scale in millimeters

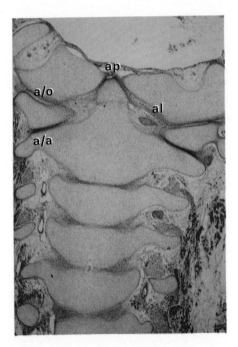

Fig. 5.3. Fetus 153N, 8 weeks, CR 30 mm. Frontal section. In this section one can clearly see: 1. The atlanto-occipital joints, showing a three-layered arrangement (*a/o*) 2. The atlanto-axial joints, showing a homogeneous interzone (*a/a*) 3. The zone of increased cellularity in the midpart of the odontoid. 4. The proximal relationship of the tip of the odontoid to the occiput 5. The apical (*ap*) and alar (*al*) ligaments between the odontoid and the foramen magnum. H & E, ×20

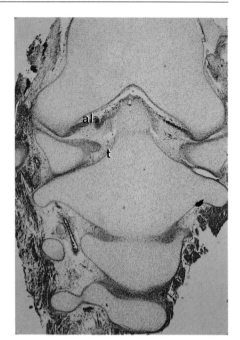

Fig. 5.4. Fetus 362, 8 weeks. Frontal section. The ligaments of the odontoid are well demonstrated in this section: 1. The alar ligaments (check ligaments), connecting the sides of the odontoid process to the foramen magnum (*al*). 2. The transverse ligaments, attached to the atlas (*t*). Azan, ×20

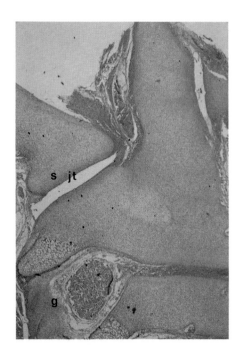

Fig. 5.5. Fetus 89M, 11 weeks, CR 70 mm. Frontal section. The ossification of the posterior elements of C-2 is well demonstrated. The synovial joint (*sjt*) between the atlas and axis is also clearly seen (including the synovial cavity). The ganglion in the foramen between C-2 and C-3 is well demonstrated (*g*). Goldner, ×20

Fig. 5.6. Fetus 206N, 12 weeks, CR 79 mm. Frontal section. This section shows multiple vascular invasions of the cervical vertebrae. Also, the two parts of the intervertebral disks: anulus fibrosus and nucleus pulposus, are well demonstrated. Goldner, × 4

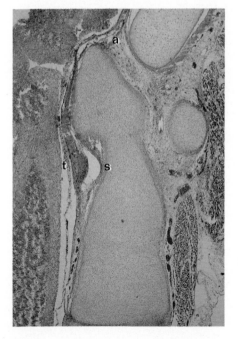

Fig. 5.7. Fetus 209M, 12 weeks, CR 78 mm. This sagittal section clearly shows the odontoid process, the body of C-2, the anterior arch of the atlas, and part of the occiput. Note: 1. The apical ligament, connecting the tip of the odontoid process to the foramen magnum (*a*). 2. The transverse ligament on the posterior aspect of the odontoid process, with its upper and lower extensions forming the cruciform ligament (*t*). 3. The synovial joint between the transverse ligament and the posterior aspect of the odontoid process (*s*). 4. The multiple blood vessels in the ligaments of the odontoid. 5. The absence of the layer of increased cellular density at the midpart of the odontoid. Goldner, × 20

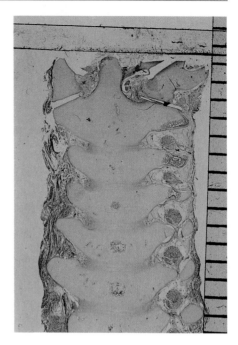

Fig. 5.8. Fetus 74M, 13 weeks, CR 90 mm. Frontal section. The well-advanced ossification in the posterior elements of C-2, C-3, and C-4 is well shown. The beginning of ossification of the bodies of the cervical vertebrae by central vascular invasion is well demonstrated. Goldner, × 10

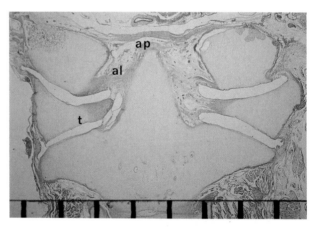

Fig. 5.9. Fetus 228M, 16 weeks, CR 130 mm. Frontal section. This section clearly shows the apical (*ap*), alar (*al*) and transverse (*t*) ligaments of the odontoid complex. The cranial position of the tip of the odontoid process in relation to the occiput is well demonstrated. The abundance of blood vessels in the apical, accessory alar, and transverse ligaments is well shown. HPS, × 10

References

1. Angevin JB (1973) Clinically relevant embryology of the vertebral column and spinal cord. Clin Neurosurg 20:95
2. Hensinger RN (1989) Congenital anomalies of the odontoid. In: Cervical Spine Research Society (ed) The cervical spine, 2nd ed. Lippincott, Philadelphia, p 248
3. Noback CR, Robertson GG (1951) Sequences of appearance of ossification centres in the human skeleton during the first five pre-natal months. Am J Anat 89(1):1
4. Ogden JA, Grogan DP (1987) The scientific basis of orthopaedics, 2nd edn. Appleton and Lange, Norwalk, pp 55–69
5. O'Rahilly R, Muller E, Meyer DB (1983) The human vertebral column at the end of the embryonic period proper. The occipitocervical region. J Anat 136:181–195
6. Schatzker J, Rorabeck CH, Waddell JP (1975) Non-union of the odontoid process. An experimental investigation. Clin Orthop 108:127
7. Sensenig EC (1949) The early development of the human vertebral column. Contrib Embryol Carnegie Inst 33:21–41
8. Sherk HH, Parke WW (1989) Developmental anatomy. In: Cervical Spine Research Society (ed) The cervical spine, 2nd edn. Lippincott, Philadelphia, pp 1–7
9. Tsou PM (1977) Embryology of congenital kyphosis. Clin Orthop 128:18–26

Chapter 6

The Development of the Thoracic Spine

H.K. Uhthoff

In general the development of the thoracic spine, and in particular its lower part, is ahead of that of either the cervical or the lumbar spine. At 6 weeks the vertebral body with its pedicles is easily identifiable as one cartilaginous anlage (Fig. 6.1). The pedicles grow in a posterolateral direction. All ribs are present as cartilaginous structures in close proximity with, but separated from, body and neural arches by blastemic tissue of uniform density. This interzone becomes more distinct at 7 weeks. By then vertebral bodies have a square shape in the sagittal plane. The relation of ribs to the vertebral bodies can be nicely seen at 8 weeks (Fig. 6.3). The diameter in the sagittal plane of vertebrae is always smaller than that in the frontal plane, as is the case in the lumbar spine. When by 10 weeks vascular invasion of the vertebrae starts, and it does so mostly from the posterior aspect, enchondral ossification of ribs is well on its way (Fig. 6.6). The three-layered interzone of the costotransverse joints is now well seen (Fig. 6.6). At 10 weeks the fusion of the right and left laminae posteriorly has taken place (Fig. 6.7), a fact observed by Bardeen as early as 1905 [1]. Enchondral ossification starts first in the ribs (Fig. 6.6). Shortly after, it takes place in the middle of the body and in both pedicles. Usually ossification is more advanced in the pedicles than in the body. There is only one ossification center in each body. The ossific nucleus increases in size at 12 weeks (Fig. 6.9) and cavitation of the costovertebral and costotransverse joints takes place. At 14 weeks the ossific nucleus nearly occupies the entire vertebra in the sagittal plane (Fig. 6.10). The size of vessels streaming in from the posterior aspect is impressive. The lateral expansion of the ossific nucleus is much slower and far from being complete at 20 weeks (Fig. 6.14) when compared with its anteroposterior extension (Fig. 6.16).

The intervertebral disk is composed of cells of uniform density at 6 weeks. The very anterior border seems slightly denser, indicating the early presence of the anulus fibrosus (Fig. 6.1). By 7 weeks the disk is of regular height (Fig. 6.2), but 1 week later a profound change has taken place (Fig. 6.3). The anulus fibrosus now forms a ring at the periphery, as its name indicates. At the center, neighboring bodies seem to touch. Closer inspection, however, reveals a thin layer of cells different from those of the bodies. Contrary to chondrocytes, these cells have an oblong nucleus and they do not sit in lacunae. Moreover, the presence of notochordal cells helps to dispel the erroneous impression that neighboring bodies touch. Inside the vertebrae the only remnant of the notochord, the mucous streak, can be recognized. At 10 weeks perinotochordal cells can be nicely observed on transverse sections. Their structure is different from prechondrocytes, a fact confirmed by their future development. During subsequent growth notochordal cells will proliferate in the disk space. This cellular agglomeration will become one part of the nucleus pulposus (Figs. 6.10, 6.12, 6.13, 6.14, 6.15, and 6.16). The other part consists of surrounding perinotochordal cells which assume a very loose quasimyxoid appearance. Their matrix stains very weakly as compared with that of the notochordal cells. The oblique direction of the outer fibers of the anulus fibrosus crossing each other at 90° is identical to that seen in the lumbar spine (see Chap. 7, Fig. 7.10).

Cavitation of facet joints starts at 7½ weeks (Fig. 6.5) and it is complete by 10 weeks. At that time the pars interarticularis is ossifying. Anterior and posterior longitudinal ligaments can be recognized at 7 weeks (Fig. 6.2). They are of identical thickness.

Cells occupying the area of the neural tube at 6 weeks are not structured in a way that permits recognition of the future spinal cord (Fig. 6.1). Nerve roots can be observed at 7 weeks. A tremendous change has occurred at 10 weeks when the spinal cord with its anterior and posterior horns, the spinal ganglia, and nerve roots remind the observer of sections taken from persons in later life (Figs. 6.7 and 6.8). Even sympathetic ganglia in front of the vertebral bodies can be recognized (Fig. 6.7). The relatively large size of the intervertebral foramen, easily accommodating spinal ganglion and nerve root, can be appreciated at 16 weeks (Fig. 6.11). Very loose connective tissue and blood vessels surround the nerve structures.

Although not belonging to the thoracic spine the sternum will be briefly mentioned on account of its interesting development. As can be seen in Fig. 6.4 it develops as a structure which at this stage of 7½ weeks is bipartite. It will fuse during subsequent growth. Absence of fusion will lead either to sternal fissures or sternal foramina.

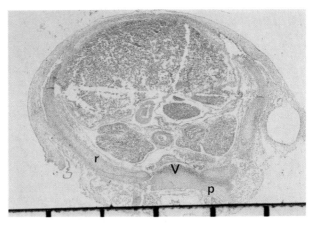

Fig. 6.1. Embryo 117N, stage 18, 6 weeks, 14 mm. Transverse section of the entire body of the embryo. Already at this stage the cartilaginous anlagen of all vertebral bodies (*V*) and ribs (*r*) are formed. Ribs are separated from the bodies by loose-celled tissue. Note that the pedicles (*p*) and the vertebral body form one cartilaginous anlage; the pedicles grow in a posterolateral direction. Goldner, scale in millimeters

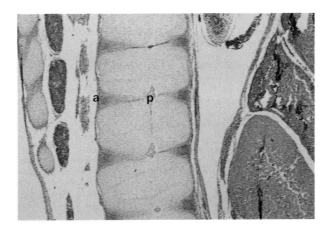

Fig. 6.2. Embryo 195N, stage 21, 7 weeks, 20 mm. This sagittal section shows well the intervertebral disks. At their periphery the beginning of the formation of the anulus fibrosus (*a*) can be suspected. Remnants of the notochord, the future nucleus pulposus (*p*), are easily identifiable. Both the anterior and the posterior longitudinal ligaments are well seen. The posterior ligament is separated from the body by loose-celled tissue. Chondrocytes of the vertebral bodies are in Streeter phase 3. Goldner, × 40

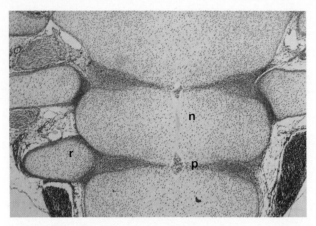

Fig. 6.3. Embryo 153N, stage 23, 8 weeks, 30 mm. Frontal section. The anulus fibrosus is now well formed. Only a narrow layer of cells remains in the midpart of the disk space. In its center proliferating notochordal cells (*p*) are seen. The mucous streak (*m*) is the only remnant of the notochord inside the vertebral body. Chondrocytes have progressed to Streeter phase 4. The tenth to the twelfth ribs (*r*) are seen and the twelfth rib is separated from the neighboring bodies by an interzone of uniform density. Below it, the spinal ganglion can be seen. HPS, × 40

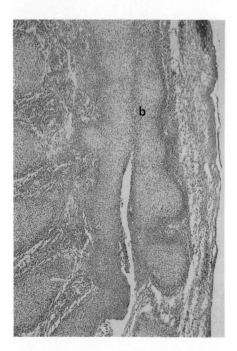

Fig. 6.4. Embryo E5, stage 21, 7½ weeks, 24 mm. Frontal section slightly oblique toward the sagittal plane. The sternum is formed from two separate cartilaginous anlagen. In its distal part, the two halves are still separated whereas a narrow band (*b*) of densely packed cells further proximally indicates the site of fusion. Goldner, × 40

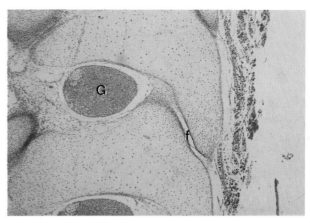

Fig. 6.5. Embryo 193N, stage 23, 7½ weeks, 25 mm. Sagittal section. Cavitation of the facet joint (*f*) is starting. Remnants of the chondrogenous layers of the interzone account for the cell density at the joint surfaces. Note the ganglion (*G*) in the intervertebral foramen. Goldner, × 40

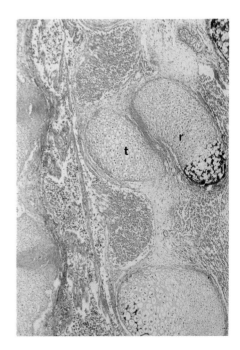

Fig. 6.6. Fetus 69, 10 weeks, 50 mm. Sagittal section. This section, taken at the level of T-12, shows an early enchondral ossification of the twelfth rib (*r*), proving that ossification occurs earlier in ribs than in vertebral bodies or pedicles. The interzone of the costotransverse joint shows a uniform cell density. Vascular invasion of the vertebral bodies is seen, starting from the posterior aspect. Transverse process (*t*). HPS, × 40

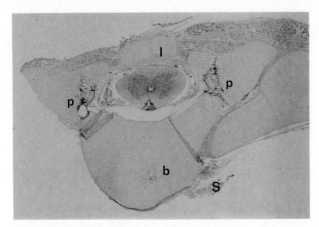

Fig. 6.7. Fetus E9, 10½ weeks, 60 mm. Transverse section. Advancing ossification in the pedicles (*p*) and beginning ossification in the middle of the body (*b*) is seen. Both processes lag behind that of the rib; the latter is separated from the body and the transverse process by a three-layered interzone. Laminae have joined posteriorly (*l*). Note also the well-formed spinal cord. Sympathetic ganglion (*S*). Goldner, same magnification as Fig. 6.1

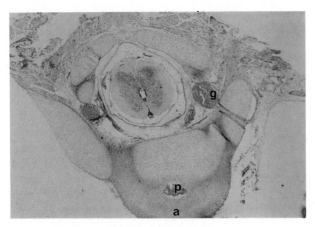

Fig. 6.8. Same fetus as Fig. 6.7. This transverse section crosses the spinal column at the sixth intervertebral disk space and shows well the anterior part of the anulus fibrosus (*a*) as well as the proliferating notochordal cells (*p*). The spinal cord and two spinal ganglia (*g*) can be well seen. HPS, same magnification as Fig. 6.1

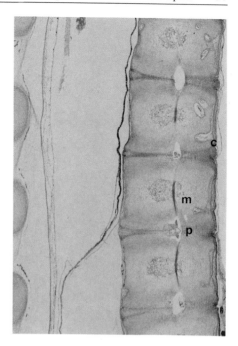

Fig. 6.9. Fetus 109N, 12 weeks, 78 mm. Frontal section. The first phase of enchondral bone formation inside the vertebra is nicely shown and the mucous streak (*m*) stains very well. Note also the presence of cartilage canals in corners of the body (*c*). The notochordal cells continue to proliferate (*p*), the remaining part of the disk space is not too distinct. Azan, same magnification as Fig. 6.1

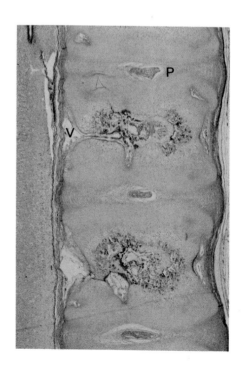

Fig. 6.10. Fetus 107N, 14 weeks, 105 mm. Sagittal section. Enchondral bone formation progresses much faster in the anterior and posterior directions than laterally. The vascular pedicle (*V*) posteriorly is well seen. Here the ligament is not attached to the body. The notochordal cells spread horizontally. Together with the surrounding loose-celled tissue they form the nucleus pulposus (*p*). Goldner, same magnification as Fig. 6.1

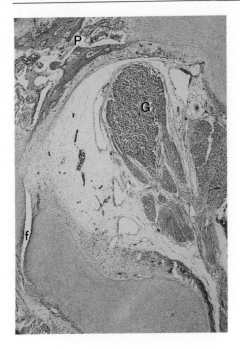

Fig. 6.11. Fetus 99N, 16 weeks, 135 mm. Sagittal section. Ganglion (*G*) and nerve root occupy the intervertebral foramen. They are surrounded by loose connective tissue. The pedicle (*P*) is ossified, and is seen to continue into the pars interarticularis. The inferior facet joints (*f*) are well formed. Azan, same magnification as Fig. 6.1

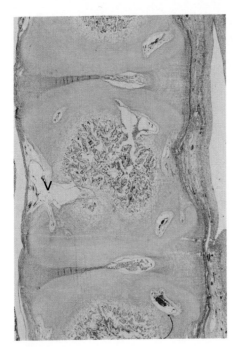

Fig. 6.12. Same fetus as Fig. 6.11. This sagittal section shows the progressing ossification of a vertebral body. A large blood vessel (*V*) enters from the posterior aspect. The nucleus pulposus and the anulus fibrosus are well seen. Azan, same magnification as Fig. 6.1

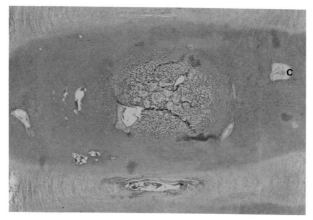

Fig. 6.13. Fetus 10, 18 weeks, 165 mm. This frontal section shows again the slow lateral progression of ossification. Note the cartilage canals in all four corners (*c*). Nucleus pulposus and anulus fibrosus are well seen. Toluidine blue, same magnification as Fig. 6.1

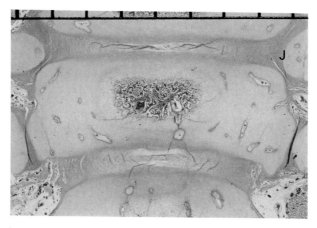

Fig. 6.14. Fetus 182N, 20 weeks, 185 mm. Frontal section. Cavitation of the costovertebral joints (*J*) is obvious. There is little progression of vertebral ossification in a lateral direction. The anulus fibrosus is easily identifiable. Note the multitude of cartilage canals. Azan, scale in millimeters

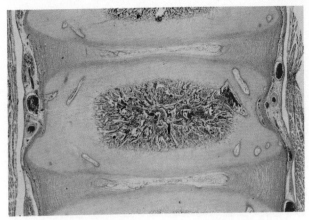

Fig. 6.15. Same fetus as Fig. 6.14. This frontal section goes more through the midportion of the body. Columns of chondrocytes are well seen; they represent the second phase of enchondral bone formation, the growth plate. Note the extent and the composition of the nucleus pulposus. Azan, same magnification as Fig. 6.14

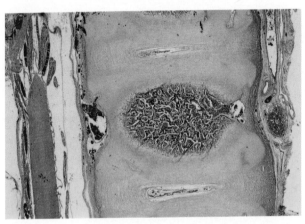

Fig. 6.16. Fetus 1, 20 weeks, 185 mm. Sagittal section. Ossification of the body has reached its anterior and posterior wall. Notochordal cells seem less compacted; their matrix contains a lightly staining material. They are surrounded by loosely arranged perinotochordal cells. The content of collagen in the anulus fibrosus has gradually increased over the weeks, changing the collagen/cell ratio and resulting in a lighter staining. Both longitudinal ligaments are well seen. Toluidine blue, same magnification as Fig. 6.14

References

1. Bardeen CR (1905) The development of thoracic vertebrae in man. Am J Anat 4:163–174
2. Sensenig EC (1949) The early development of the human vertebral column. Contrib Embryol Carnegie Inst 33:21–41

Chapter 7

The Development of the Lumbar Spine

H.K. UHTHOFF and H.H. EL-KOMMOS

At the age of 6 weeks when the embryo measures 14 mm in length, vertebral body and pedicles are forming a single anlage (Fig. 7.1). A rather rounded shape of the vertebrae seen in the sagittal plane changes during the next few days into a squarer form, more obvious at 7 weeks (Fig. 7.2). The vertebral diameter in the sagittal plane (Fig. 7.3) is always smaller than that in the frontal plane. Progressing chondrocyte maturation of vertebrae and pedicles is seen at 8 weeks. At that stage the mucous streak inside the vertebral bodies is the only remnant of the notochord. Here, most notochordal cells have undergone degeneration. Vascular invasion of the vertebral bodies is seen at 10 weeks and seems to start from the posterior aspect. Chondrocytes in the center of the vertebral bodies and in the pedicles are now in Streeter phase 4. By 12 weeks the vessels have reached the center of the vertebrae where chondrocytes have now degenerated (Streeter phase 5). Thus, enchondral ossification of the lumbar spine starts later than that of the thoracic spine; this first phase of enchondral ossification is starting at three sites simultaneously, that is in the body and in both pedicles. By 14 weeks progression of ossification at all three sites is noted. Cartilage canals are present in all four corners of the vertebra, but no ossification takes place here (Fig. 7.15). Also there is no evidence of chondrocyte hypertrophy. Enchondral bone formation of the vertebral bodies progresses faster in the sagittal plane than in the frontal plane. The center of ossification increases in size with age (Fig. 7.15 and 7.16). By 20 weeks it occupies the entire depth of the body (Fig. 7.17), but both lateral walls still consist of cartilage. Columns of chondrocytes, cranial and caudal to the center of ossification, indicate the beginning of the second phase of enchondral ossification (see Chap. 2). Vascular ingrowth throughout the entire period is always more obvious from the posterior side.

The pedicles, already seen at 6 weeks, grow in a posterolateral direction, giving an impression of splaying. Soft tissues cover the neural tube posteriorly. With progression of growth of the posterior elements, facet joints are seen as uniform blastemic condensations (Fig. 7.2) between the articular processes; a three-layered interzone of the facet joints becomes visible at 7½ weeks (Fig. 7.5). At 8 weeks, the neural arches advance in a more posterior direction and by 10 weeks the tips of the laminae grow toward the midline (Fig. 7.7). However, at that time they are not fused, but united by a band of fibrous tissue. At 10½ weeks, fusion of both laminae has occurred (Fig. 7.10a, b). By 14 weeks facet joints are well formed and ossification is seen in the pars interarticularis (Fig. 7.13).

Although a dense cellular zone between developing vertebral bodies, indicating the site of the intervertebral disk, is already visible at 6 weeks, no cellular differentiation is recognizable. Notochordal cells are seen inside the future disk. The dense cellular area narrows during the next few days and the formation of the anulus fibrosus starts (Fig. 7.2). At 7 weeks the ring-like structure of the anulus becomes apparent (Fig. 7.3). At the midportion of the disk space, neighboring vertebral bodies are separated by a narrow layer of densely packed cells (Fig. 7.4). Transverse sections of the area around the notochord show cells resembling prechondrocytes (Streeter phase 1; Fig. 7.3). They are known as perinotochordal cells. These cells are less densely packed than those of the anulus. Notochordal cells at the level of the disks proliferate whereas they degenerate inside the bodies (Fig. 7.4). At 8 weeks the anulus is more evident and elaboration of collagen fibers helps to heighten its well-known structural aspect (Fig. 7.6). Figure 7.6 reveals well the wedge shape of the anulus. The circular arrangement of its fibers can be appreciated on transverse sections (Fig. 7.10a). An interesting fiber orientation of the anterior part of the anulus fibrosus can be observed in frontal sections (Fig. 7.11). These fibers take an oblique direction and cross each other at 90°. The notochordal cells in the disk space begin to proliferate and grow laterally into the loose-celled area, well seen at 12 and more so at 14 weeks (Fig. 7.12). Now we can speak of a nucleus pulposus. Remnants of the notochord inside the body can be seen as a vertical, deeply stained band, the mucous streak (Fig. 7.8). In sections going exactly through the midline one gains the wrong impression that this streak divides the body into two halves. The notochordal and perinotochordal cells form the nucleus pulposus (Fig. 7.12). During subsequent development the disk space increases in height.

The formation of ligaments becomes obvious at 8 weeks when both the posterior and the anterior longitudinal ligaments can be seen (Fig. 7.6). Both are attached to the anulus fibrosus. Loose connective tissue, richly vascularized, separates the posterior longitudinal ligament from the posterior vertebral wall. The interspinous and particularly the strong supraspinous ligament which is less cellular, but more collagenous, are well formed at 14 weeks (Fig. 7.14). Obviously at that time the spinous processes are cartilaginous structures.

At 6 weeks the neural tube consists of a conspicuously large ventricle surrounded by a germinal layer (Fig. 7.1). At 7 weeks ganglia and nerve roots can be recognized (Fig. 7.2) and their relative size is worth noting. Figure 7.5 taken at 7½ weeks shows the intervertebral foramen containing ganglion and nerve root. At 10 weeks the spinal cord has assumed its familiar shape and structure (Fig. 7.7). Ganglia can be seen inside the intervertebral foramina (Fig. 7.7). Both structures are also well seen at the sacral level (Fig. 7.9), where the nerve root exits anteriorly.

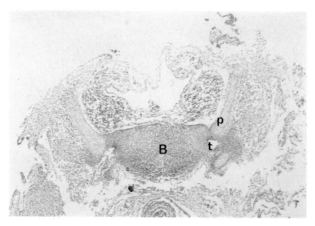

Fig. 7.1. Embryo 117N, stage 18, 6 weeks, 14 mm. Transverse section. The section goes through the level of L-1 and shows that the posterior elements (*p*) are in continuity with the body (*B*); they grow in a posterolateral direction. Soft tissues are closing the neural tube posteriorly. Note the rather small notochord. The transverse process (*t*) of one side is well seen. Goldner, × 16

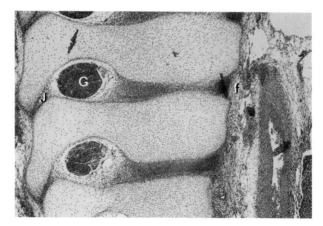

Fig. 7.2. Embryo 150N, stage 19, 7 weeks, 18 mm. Sagittal section. The anulus fibrosus (*f*) is now identifiable. An interzone of uniform cellular density is seen at the facet joints (*J*). Rather huge ganglia (*G*) occupy the foramina. Goldner, × 40

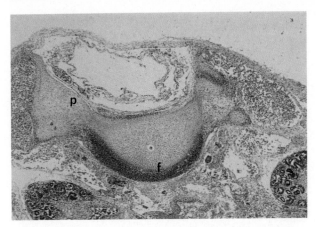

Fig. 7.3. Embryo E17, stage 20, 7 weeks, 20 mm. Chondrocytes of the posterior elements (*p*) are in Streeter phase 4. The anulus fibrosus (*f*) is easy to identify at the anterior aspect of the body, but less well seen posteriorly. Precartilage-like cells surround the notochord. Azan, × 16

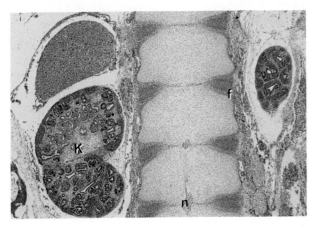

Fig. 7.4. Embryo 119N, stage 19, 7 weeks, 20 mm. Frontal section. Vertebral bodies are separated at the periphery by a well-defined anulus fibrosus (*f*). More centrally, a narrow strip of tissue slightly denser than the surrounding chondrocytes forming the intervertebral disk is seen between the bodies. At the midpart, the proliferation of notochordal cells is obvious (*n*). Kidney (*K*). Goldner, × 16

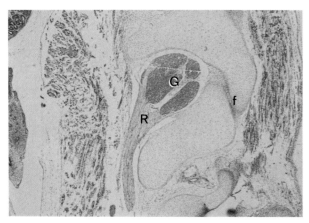

Fig. 7.5. Embryo 193N, stage 23, 7½ weeks, 25 mm. Sagittal section. The ganglion (*G*), and nerve root (*R*) of L-3 are well demonstrated. The facet joint (*f*) shows a three-layered arrangement. HPS, × 40

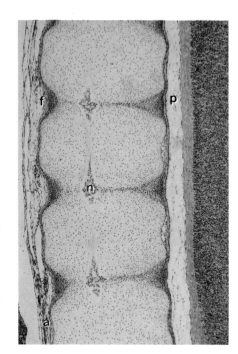

Fig. 7.6. Embryo 151N, stage 23, 8 weeks, 27 mm. Sagittal section. This section, not exactly through the midline of the spine, shows the anulus fibrosus (*f*) and a narrow strip of cells constituting the intervertebral disk. Cells of the notochord (*n*) proliferate at the level of the intervertebral disk. The anterior longitudinal ligament (*a*) is attached to body and disk; the posterior longitudinal ligament (*p*) is separated from the body by loose connective tissue. Goldner, × 40

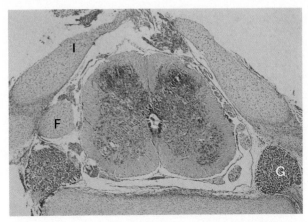

Fig. 7.7. Fetus E9, 10½ weeks, 60 mm. Transverse section. This section of L-3 shows that the laminae (*l*) have not yet fused although they grow toward each other. At one side the superior facet of L-4 (*F*) is seen. Spinal ganglia (*G*) are situated in both foramina and the spinal cord is well formed. HPS, ×16

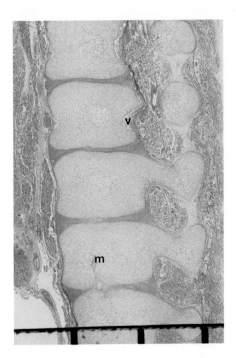

Fig. 7.8. Fetus 69, 10 weeks, 50 mm. Sagittal section. Inside L-3 a mucous streak (*m*) containing a few degenerated notochordal cells is seen. Chondrocytes in body and pedicles are in Streeter phase 4 and vascular invasion (*v*) of the bodies at their posterior aspect is beginning. HPS, scale in millimeters

Fig. 7.10a, b. Fetus 379, 10½ weeks, 56 mm, horizontal section. **a** Section at the level of S-3 showing nicely the anulus fibrosus (*a*). The posterior elements are still not fused. Ilium (*I*); sacral ala (*S*). **b** Section at the level of L-4. The posterior elements are fusing (*F*). The spinal cord is still present at this level. Azan, scale in millimeters

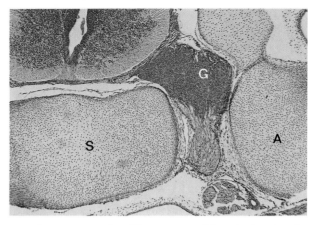

Fig. 7.9. Fetus E6, 10 weeks, 50 mm. Transverse section. The structure of the spinal cord at the level of the sacrum (*S*) is nicely shown. Note also the spinal ganglion (*G*) and the nerve root which exits anteriorly. Sacral ala (*A*). Goldner, ×40

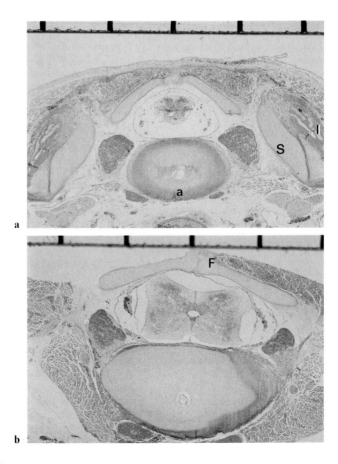

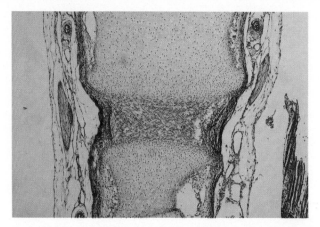

Fig. 7.11. Fetus 194, 12 weeks, 75 mm. This frontal section goes through the anterior fibers of the anulus fibrosus and shows oblique fibers crossing each other at 90°. Azan, × 40

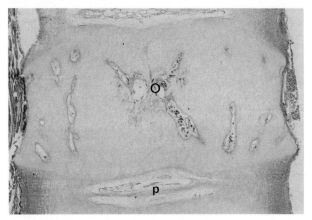

Fig. 7.12. Fetus 23N, 14 weeks, 105 mm. Frontal section. Ossification (*O*) starting inside the third lumbar vertebra. Note also the expansion of notochordal cells which together with the loosely arranged surrounding cells form the nucleus pulposus (*p*). Azan, same magnification as Fig. 7.8

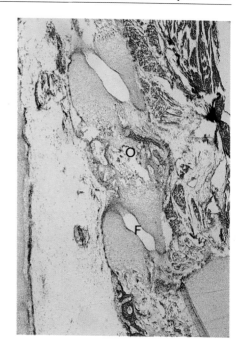

Fig. 7.13. Fetus 82N, 14 weeks, 105 mm. Oblique longitudinal section. Ossification (*O*) of the pars interarticularis is seen as well as the facet joints (*F*). Azan, ×20

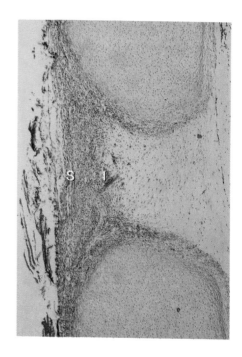

Fig. 7.14. Embryo 143N, 14 weeks, 110 mm. Sagittal section. This microphotograph shows the supraspinous ligament (*S*) containing well-oriented collagen fibers. The interspinous ligament (*I*) is less well organized and contains more cells. The spinous processes are still cartilaginous. HPS, ×40

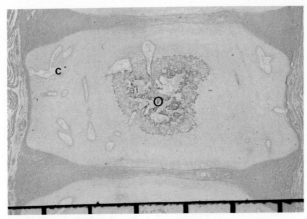

Fig. 7.15. Fetus 25, 16½ weeks, 140 mm. Frontal section. This shows the extent of the ossific nucleus (*O*) which is far from reaching the lateral walls of the vertebral body. Of note are the cartilage canals (*c*). HPS, scale in millimeters

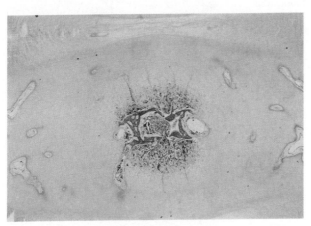

Fig. 7.16. Fetus 10, 18 weeks, 165 mm. Frontal section. There is a slow progression of ossification in the frontal plane. Although the cartilage canals in all four corners of the vertebra are getting bigger, no ossification occurs at these sites. Azan, same magnification as Fig. 7.15

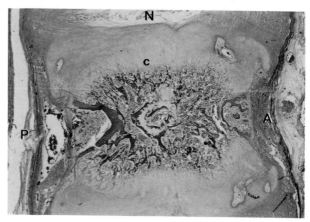

Fig. 7.17. Fetus 115N, 20 weeks, 185 mm. Sagittal section. The center of ossification occupies the entire depth of the vertebral body. Hematopoietic cells are seen between the trabeculae. Chondrocytes distal and proximal to the ossific nucleus line up in columns (*c*). Note the development of the disk and the nucleus pulposus (*N*) in the upper part of the microphotograph. Here, loosely arranged notochordal cells are surrounded by loose cellular mucoid-like tissue, both forming the nucleus pulposus. Anterior longitudinal ligament (*A*); posterior longitudinal ligament (*P*). Azan, same magnification as Fig. 7.15

References

1. Goto S, Uhthoff HK (1985) Notochord action on spinal development: a histologic and morphometric investigation. Acta Orthop Scand 57:85–90
2. O'Rahilly R, Benson DR (1985) The development of the vertebral column. In: Bradford DS, Hensinger RM (eds) The pediatric spine. Stuttgart, Thieme
3. O'Rahilly R, Muller F, Meyer DB (1980) The human vertebral column at the end of the embryonic period proper. J Anat 131:565–575
4. Peacock A (1951) Observations on the prenatal development of the intervertebral disc in man. J Anat 85:260–275
5. Sensenig EC (1949) The early development of the human vertebral column. Contrib Embryol Carnegie Inst 33:21–41
6. Tanaka T, Uhthoff HK (1981) Significance of resegmentation in the pathogenesis of vertebral body malformation. Acta Orthop Scand 52:331–338

Chapter 8

The Development of the Shoulder

H.K. UHTHOFF

In order to ease the understanding of the illustrations, frontal sections have been chosen and only occasionally have transverse cuts been used. As may be expected, some deviation from the exact plane occurred during embedding and sectioning which sometimes makes the interpretation of findings difficult.

The first sign of development of the shoulder joint is seen at 6 weeks when the humeral anlage is separated from the cartilaginous glenoid by a uniform interzone. The acromion also shows early cartilage formation (Fig. 8.1). The tremendous speed of development is obvious in Fig. 8.2, taken at 6½ weeks, where one sees the humerus and scapula separated by a three-layered interzone. The beginning of glenohumeral cavitation is seen as a further loosening of the middle layer of the interzone at 7 weeks. A more advanced cavitation is seen at 8 weeks when the embryo measures 30 mm (Fig. 8.4). A complete joint cavity lined with synovial tissue, however, will only be seen at 10 weeks (Fig. 8.6). Evidence of cavitation of the glenohumeral joint has been observed by Gardner and Gray [1] in an embryo measuring 22 mm (7½ weeks). A joint space is always present in its entirety although in some cases, owing to specimen preparation, it cannot be fully visualized.

At the site of the future acromioclavicular joint, condensation of cells of the acromioclavicular interzone is well seen at 6.5 weeks. At 10 weeks the acromioclavicular joint cavity becomes obvious. The formation of the clavicle starts with two separate centers of membranous (direct) ossification (see Chap. 15, Fig. 15.10). Ossification of the lateral part is well seen in Fig. 8.2. Both centers will fuse. This is followed by the appearance of cartilage at both ends. The latter are the site of enchondral ossification (epiphyseal growth plate) which assures the growth in length (Fig. 8.5; see Chap. 2). The growth of the clavicle at its acromial end, which is less rapid than that of the medial part, is well shown in Figs. 8.6, 8.8, and 8.9.

At stage 17 a cellular condensation, starting at the superior portion of the glenoid and extending over the humeral head, represents the long head of the biceps tendon, as previously described by Gardner and Gray [1]. At a later stage its intra-articular presence is easily recognized (Fig. 8.10).

The formation of the labrum starts at stage 19. At that time the attachment of the deltoid to the acromion can also be seen (Fig. 8.3). The labrum becomes very obvious at 10 weeks as a dense structure (Fig. 8.6). At the inferior aspect of the glenohumeral joint, the capsule inserts into the labrum. With time the inferior capsule will become redundant and will form a recessus, well seen in Fig. 8.9. At the anterior aspect, however, the capsule does not always attach to the labrum. In our series of specimens older than 12 weeks we observed that the anterior capsule attaches to the labrum in only 50% of cases [3]. In the remaining instances the capsule inserts into the neck of the scapula (Fig. 8.10). The frequent attachment of the anterior capsule into the neck of the scapula instead of into the labrum should not be interpreted as a Bankart lesion which is a true detachment of the labrum from the glenoid rim. The posterior capsule, however, always inserts into the labrum.

Development of ligaments, tendons, and muscles can already be seen at 7 weeks when the embryo measures between 18 and 21 mm. All rotator cuff muscles are present. The spine of the scapula develops and some ligaments can be suspected by the presence of more cellular streaks. Neurovascular structures below the glenohumeral joint are obvious. According to Gardner and Gray [1], the "form of elements" resemble that of an adult shoulder once an embryo has reached a length of 22 mm (7.5 weeks). This seems slightly exaggerated, particularly in respect of the relative size of the different elements and the absence of the metaphyseal flare. At 10 weeks a distinction can be made between supraspinous muscle and capsule (Fig. 8.6); it is more obvious at 16 weeks (Fig. 8.9). Vascular invasion of the humeral head as well as a dense capsule are recognizable at 10 weeks (Fig. 8.6). The transition between muscles and tendons is abrupt at 12 weeks (Fig. 8.7) whereas it is more gradual at 20 weeks (Fig. 8.11). The vascular supply to the supraspinous tendon is well seen; more vessels are present in the subbursal tissues than in the tendon itself (Fig. 8.12). At 14 weeks vessels in the humeral epiphysis are close to the metaphyseal growth plate, but evidence of their crossing the plate is only seen 2 weeks later. We note with interest the attachment of muscles and tendons to bone at 12.5 weeks (Fig. 8.7). Here it is evident that in general they are continuous with the fibrous layer of the perichondrium or periosteum, but not with the cartilage itself (Fig. 8.7). Sharpey's fibers have not yet developed.

Small spaces representing the future subacromial bursa have been described by Gardner and Gray [1] in an embryo 22 mm in length (7.5 weeks). Formation of this bursa was seen by us only at later stages, with the exception of an isolated instance present in an embryo 27 mm in length (8 weeks). The presence of a well-defined subacromial bursa lined by synovial cells is obvious at 12 weeks (Fig. 8.7). At 20 weeks the subacromial bursa extends further laterally under the deltoid. The combined joint capsule – tendon complex of the supraspinatus forms a strong superior restraint (Fig. 8.11). At this time this restraint is more cellular than fibrous which is not the case later in life. Already at 10 weeks, the notch for the suprascapular artery and nerve is well formed (Fig. 8.6).

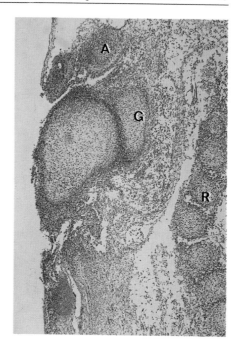

Fig. 8.1. Embryo 83, stage 17, 6 weeks, 11 mm. Frontal section. The anlage of humeral head and glenoid (*G*) is well seen. They are separated from each other by a uniform interzone. The acromion (*A*) begins to chondrify. The cleft seen in the subacromial space is an artifact. At the right border, cartilaginous anlagen of ribs can be seen (*R*). HPS, × 40

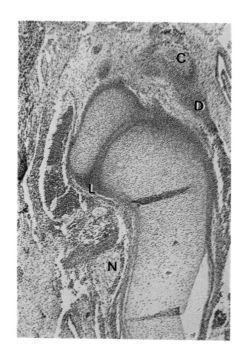

Fig. 8.2. Embryo 47N, stage 19, 6½ weeks, 15 mm. Frontal section. A three-layered interzone between humeral head and glenoid is obvious. Cellular condensation indicates the beginning of the inferior labrum (*L*). The clavicle (*C*) is ossifying and the deltoid (*D*) can be seen, but not the acromion. The neural structure medial to the humerus is the median nerve (*N*). Azan, × 40

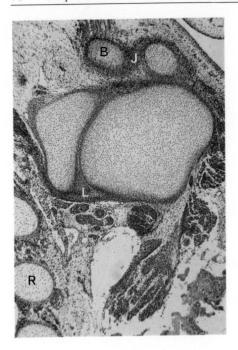

Fig. 8.3. Embryo 150N, stage 19, 7 weeks, 18 mm. Frontal slightly oblique section. The humerus is in flexion. The inferior labrum (*L*) is clearly defined. The deltoid gains in volume and a cellular condensation takes place between clavicle and acromion (*J*). Bone formation (*B*) in the clavicle stains green. Ribs are seen to the left (*R*). Goldner, ×40

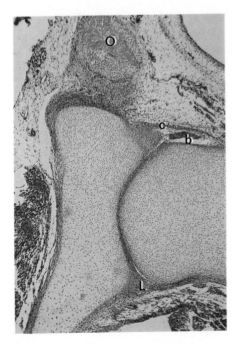

Fig. 8.4. Embryo 153N, stage 23, 8 weeks, 30 mm. Frontal section. The humerus is in abduction. There is definite evidence of cavitation at the periphery of the glenohumeral joint. The superior joint capsule (*c*) is strong and covers the dense long head of the biceps (*b*). Ossification (*O*) of the clavicle progresses. Inferior labrum (*L*). HPS, ×40

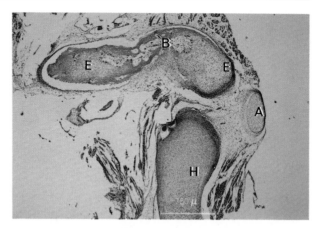

Fig. 8.5. Fetus 222N, 9 weeks, 41 mm. Frontal section. This section illustrates well the development of the clavicle. Its midpart forms through membranous bone formation (*B*) whereas enchondral bone formation at both ends assures growth in length (*E*). Note also the acromion (*A*) and the humerus (*H*). HPS, length of bar 750 μm

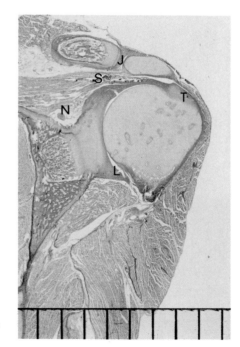

Fig. 8.6. Fetus 112N, 10 weeks, 53 mm. Frontal section. Cavitation between acromion and clavicle is evident (*J*). Only the lateral aspect of the clavicle is seen. The cartilaginous capping of the clavicle and the underlying enchondral bone formation are obvious. The subacromial bursa starts to form. The glenohumeral joint reaches further around the head, and synovial folds start to appear. The supraspinatus (*S*) is seen blending with the capsule. Vascular invasion of the humeral head is observed. Note the advancing enchondral bone formation of the scapula. Suprascapular nerve and artery can be seen in the notch (*N*). Labrum (*L*); greater tuberosity (*T*). Goldner, scale in millimeters

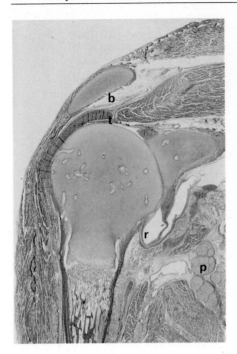

Fig. 8.7. Fetus 125N, 12½ weeks, 86 mm. Frontal section. The formation of the subacromial bursa (*b*) has taken place. The transition (*t*) from the muscular to the tendinous part of the supraspinatus is striking by its abruptness. Its tendon inserts into the perichondrium. Modeling of the humeral metaphysis is obvious. Note the inferior joint recessus (*r*). Brachial plexus (*p*). Azan, same magnification as Fig. 8.6

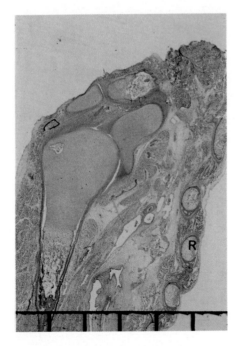

Fig. 8.8. Fetus 107N, 14 weeks, 105 mm. Frontal section. The supraspinatus and the superior joint capsule are well seen. Although more fibrous than at earlier stages, the cellular components of the capsule still predominate. The supraspinatus inserts into the greater tuberosity and collagen fibers are anchored into the cartilage. This kind of attachment seems an exception at that stage of development. Note the progression of enchondral ossification of the humerus. The inferior joint recessus is well formed. Cartilage canals are seen in the humeral head. Nerves and vessels are seen in the axilla. Ribs (*R*) are seen on the right. Goldner, scale in millimeters

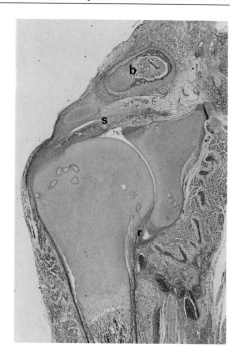

Fig. 8.9. Fetus 29N, 16 weeks, 130 mm. Frontal section. The redundant inferior capsule is well seen. Note again the enchondral bone formation (*b*) at the lateral end of the clavicle and the superior and inferior labra. Both the muscular and tendinous portions of the supraspinatus (*s*), as well as the tendon insertion into the greater tuberosity, are well identified. Inferior recessus (*r*). Azan, same magnification as Fig. 8.8

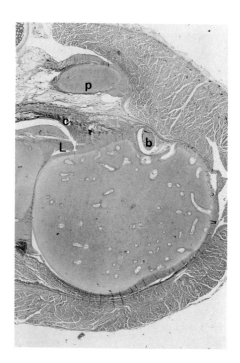

Fig. 8.10. Fetus 28N, 19 weeks, 175 mm. Transverse section. This section was chosen to show that the anterior capsule (*c*) bypasses the anterior labrum (*L*). Such an occurrence should not be confused with a Bankart lesion which is a true detachment of the labrum from the glenoid rim. The long head of the biceps (*b*) is seen in its tendon sheath and covered by the capsule. Note the coracoid process (*p*) and the thick layer of the deltoid covering the humeral head. Azan, same magnification as Fig. 8.8

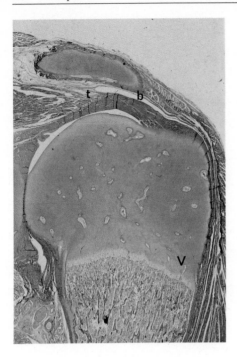

Fig. 8.11. Fetus 115N, 20 weeks, 185 mm. Frontal section. The subacromial bursa (*b*) is well seen as is the loose connective tissue further medial to it. The muscle belly of the supraspinatus and its tendinous part now show a more gradual transition (*t*). Its tendon blends with the capsule. Note one blood vessel crossing the proximal humeral physis (*V*). Azan, same magnification as Fig. 8.8

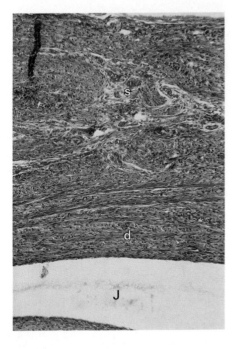

Fig. 8.12. Same fetus as in Fig. 8.11. This higher magnification shows well the supraspinous tendon. Its deeper layers (*d*) close to the glenohumeral joint (*J*) are less well vascularized than the part closer to the bursa (*s*). The spatial arrangement of tendon fibers is also more evident in the part closer to the joint. Azan, × 100

References

1. Gardner E, Gray DJ (1953) Prenatal development of the human shoulder and acromioclavicular joints. Am J Anat 92:219–276
2. Olah OJ, Ludwig KS (1971) Die Embryonalentwicklung des Sternoclaviculargelenkes beim
   Menschen. Acta Anat (Basel) 78:383–405
3. Uhthoff HK, Piscopo M (1985) Anterior capsular redundancy of the shoulder: congenital or
   traumatic? J Bone Joint Surg [Br] 67:363–366

Chapter 9

The Development of the Elbow

J. F. LOEHR and H.K. UHTHOFF

In this chapter the sagittal plane has been chosen for ease of interpretation of sequential developments. Only occasionally have we used transverse sections. The elbow is in a position of flexion with the forearm in pronation. At the age of 6 weeks flexion amounts to 60°; it progresses thereafter to reach up to 135°.

At 6 weeks the interval between humerus and radius, as well as between humerus and ulna, is occupied by homogeneous blastemic tissue. At some sites, however, a change to a three-layered formation becomes apparent (Figs. 9.1 and 9.2). Evidently, each bone has its own cartilaginous anlage and the separating blastemic tissue constitutes the site of the future joint.

During the following days a definite layering of the interzones takes place. The shape of nuclei of all three layers, however, seems identical (Fig. 9.3). Later on, nuclei of the middle layer become elongated, their main axis being parallel to the joint line. The denser outer layers are continuous with the perichondrium. Cavitation starts around the 7th week. Small cavities can be seen peripherally as well as centrally (see Chap. 3, Fig. 3.6 a, b). A similar observation was made by Gray and Gardner [1] and documented in an embryo in an identical stage of development. Later on the cavities will become confluent, but cellular strands crossing the joint will persist for some time. Early in the fetal period triangular folds start to develop (Fig. 9.5). Synovial tissue develops around the 10th week. The synovial folds are richly vascularized (Fig. 9.15).

As early as 7 weeks the first phase of enchondral bone formation of the ulna and radius is progressing (Fig. 9.4) and at 14 weeks has reached the level of the radial tuberosity (Fig. 9.12). This will be followed much later by the second phase of enchondral bone formation, the formation of the physes (see Chap. 2). At this stage we start to recognize columns of chondrocytes which constitute the beginning of the growth plate (Fig. 9.18). At 12 weeks the formation of the coronoid and olecranon fossae takes place. Figure 9.13, taken at 14 weeks, shows the deepening of both fossae. Whether this deepening is secondary to joint motion is more than doubtful.

The diameters of the radial head and neck are identical at 7 weeks (Fig. 9.4); 4 weeks later their relative size has changed and the diameter of the neck is much smaller than that of the head (Fig. 9.6). The annular ligament is seen over the radial head, extending to the neck without solid attachment to the periosteum. This arrangement can also be observed in older specimens (Figs. 9.9, 9.12, 9.14, 9.15, 9.16, and 9.18). One can project that this weak transition distally between annular ligament and surrounding structures may in some way constitute the pathogenetic mechanism of a pulled elbow [2].

The development of the joint between humerus and ulna as well as that of the coronoid and olecranon processes are well seen in Figs. 9.7, 9.8, 9.10, 9.11 and 9.17.

Vascular canals in the epiphyses can be seen at 12 weeks and become more obvious during the subsequent weeks. Their number also increases with time. At 20 weeks the maturation has progressed further without showing any major change (Fig. 9.18).

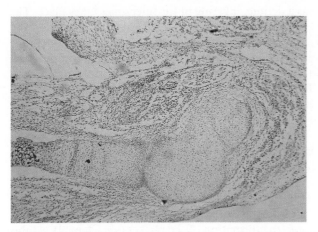

Fig. 9.1. Embryo 86N, stage 18, 6 weeks, 13 mm. Sagittal section. The lower end of the humerus as well as the radial head and neck are seen. Note that the diameters of the radial head and neck are equal. A joint capsule has not formed yet. Muscles are forming, but their identification is difficult. Note the hypertrophy of chondrocytes (Streeter phase 4) in the radial shaft. Goldener, × 40

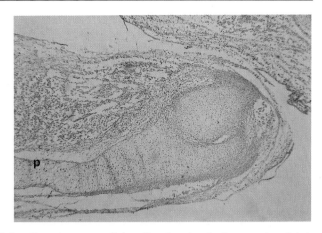

Fig. 9.2. Same embryo. Sagittal section. A more medial section showing the humeroulnar joint. The anteroposterior diameter of the trochlea is greater than that of the more proximal humeral shaft. Fibers of the brachialis muscle inserting into the ulna distal to the coronoid process, and of the triceps muscle inserting into the perichondrium of the ulna can be suspected. Beginning of a three-layered structure of the interzone. Early periosteal intramembranous bone formation of the ulnar diaphysis (*p*) can be seen. Goldner, × 40

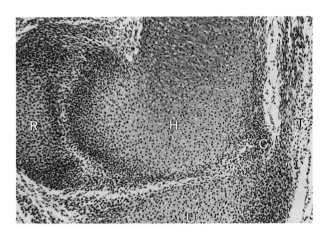

Fig. 9.3. Embryo 47N, stage 18, 6 ½ weeks, 15 mm. Sagittal section. The interzone now shows a definite three-layered appearance, especially well seen at the radiohumeral joint. Nuclei of cells in the middle layer are still round. The cleft between humerus and ulna is due to processing. The formation of a joint capsule can be suspected by a cellular condensation. Radius (*R*); ulnar (*U*); humerus (*H*); joint capsule (*C*); triceps (*T*). Azan, × 100

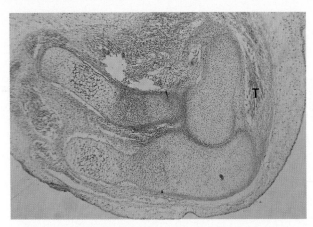

Fig. 9.4. Embryo 123N, stage 19, 7 weeks, 19 mm. Sagittal section. The triceps muscle (*T*) is more definite. Loose-celled tissue occupies the area proximal to the olecranon. No cavitation is evident yet. A periosteal bony sleeve is seen around ulna and radius. Phase 1 of enchondral bone formation is well seen in radius and ulna. Goldner, × 40

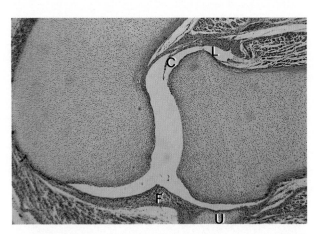

Fig. 9.5. Fetus 49N, 8½ weeks, 34 mm. Sagittal section. The shape of the radial head resembles its adult form. At its upper face the annular ligament (*L*) is seen as well as its weak distal attachment. Proximally it blends into the capsule (*C*). The articulation with the ulna (*U*) is seen in the lower part of the picture. A synovial fold (*F*) is forming. Goldner, × 40

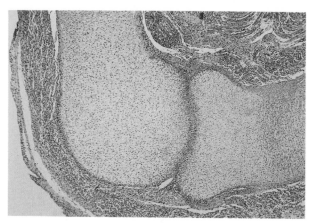

Fig. 9.6. Fetus 244NR, 10½ weeks, 57 mm. Sagittal section. The radiohumeral joint is well seen, as is part of the joint cavity. Manipulations during processing have obliterated the remaining space. The fibers of the brachialis and triceps muscles are easily recognizable. Note that the radial head is now larger than the neck. HPS, × 40

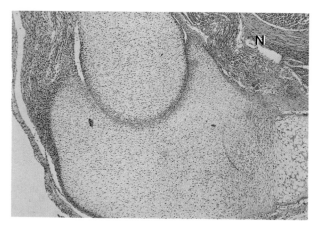

Fig. 9.7. Same fetus as in Fig. 9.6. This more medial cut shows the joint between humerus and ulna. Coronoid and olecranon processes are developing. Fibers of the brachialis muscle are seen in front of the coronoid process, and anterior to them the median nerve (*N*). HPS, × 40

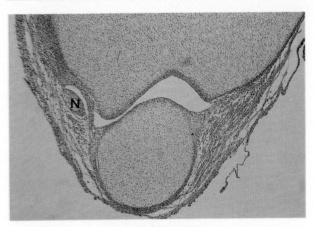

Fig. 9.8. Left elbow of same fetus. Transverse section. The articulation between olecranon and humerus is well seen. Note also the ulnar nerve (*N*). The joint capsule and the lateral expansions of the triceps can be recognized. Goldner, × 40

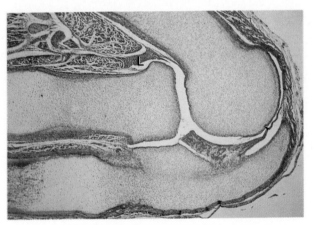

Fig. 9.9. Fetus 219N, 11 weeks, 68 mm. Sagittal section. The radiohumeral and radioulnar joints are easily recognizable. The annular ligament (*L*) is dense, it continues into the anterior capsule as a thin strand. Its weak distal attachment to the neck is again noted. HPS, × 25

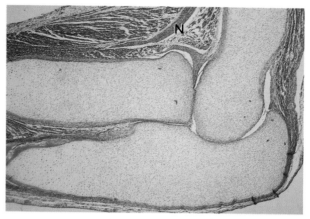

Fig. 9.10. Fetus 221NR, 12 weeks, 77 mm. Sagittal section. The triceps inserts into the fibrous perichondrium of the olecranon. A posterior humeral indentation indicates the formation of the olecranon fossa. The annular ligament is well seen. Note the radial nerve (*N*). HPS, ×25

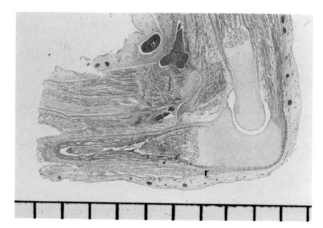

Fig. 9.11. Fetus 125NR, 12½ weeks, 80 mm. Sagittal section. This section at the level of the humeroulnar joint shows well all pertinent structures. The median nerve and brachial artery as well as the cephalic vein are seen. Note also the insertion of the triceps and the brachialis muscle. At both the humerus and the ulna, the ring of Lacroix (*r*) is well seen. Early formation of subcutaneous tissue. HPS, scale in millimeters

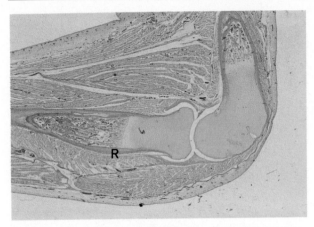

Fig. 9.12. Fetus 42NR, 14 weeks, 110 mm. Sagittal section. The relation of the annular ligament to the anterior joint capsule is well seen. At the anterior and posterior aspects of the radiohumeral joint the synovial folds increase in size. Vascular invasion of the humeral epiphysis starts. Note the presence of hematopoietic marrow in humerus and radius. Radial tuberosity (*R*). Goldner, same magnification as in Fig. 9.11

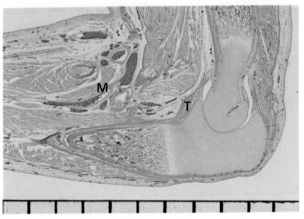

Fig. 9.13. Same fetus as in Fig. 9.12. Sagittal section. This more medial cut goes through the humeroulnar joint. The tendinous insertion of the brachialis (*T*) distal to the coronoid process is well seen. Note the increasing depth of the olecranon and coronoid fossae. The brachial artery and the superficial veins are well seen, as is the median nerve (*M*). Goldner, scale in millimeters

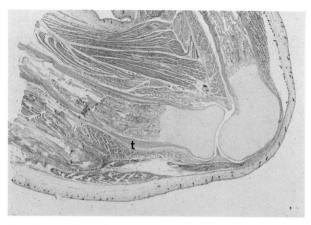

Fig. 9.14. Fetus 228NR, 16 weeks, 130 mm. Sagittal section. At this time the formation of the radial tuberosity (*t*) becomes apparent. The annular ligament is stronger over the neck than over the head of the radius. HPS, same magnification as in Fig. 9.13

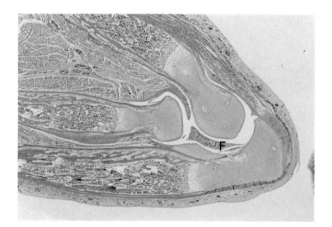

Fig. 9.15. Fetus 99NR, 16 weeks, 135 mm. Sagittal section. This section shows well the rich blood supply to the synovial tissue. The increasing thinning of the humerus at the level of the fossae is evident. Vascular invasion of all three epiphyses is seen. Synovial folds (*F*). Goldner, same magnification as in Fig. 9.13

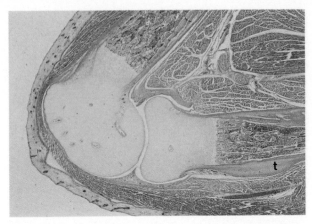

Fig. 9.16. Fetus 156NR, 17½ weeks, 154 mm. Sagittal section. The intra-articular situation of the entire radial head is well seen. Note again the radial tuberosity (*t*). HPS, same magnification as in Fig. 9.13

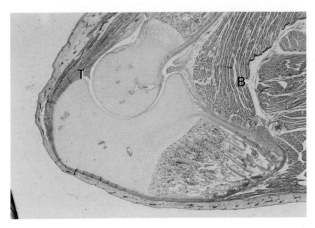

Fig. 9.17. Same fetus as in Fig. 9.16. Sagittal section. This section shows the sites of attachment of the triceps (*T*) and brachialis (*B*). In particular, the thickness of the latter muscle is impressive. The fibers of the triceps are in continuity with the perichondrium. HPS, same magnification as in Fig. 9.13

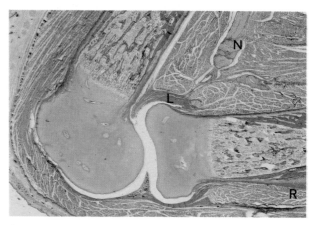

Fig. 9.18. Fetus 115NR, 20 weeks, 185 mm. Sagittal section. This section shows again that the annular ligament (*L*) is not as strong directly over the radial head as at a slightly more distal site. The radial tuberosity (*R*) is seen again posteriorly, indicating that the forearm is in pronation. The radial nerve is sectioned obliquely (*N*). Azan, same magnification as in Fig. 9.13

References

1. Gray DJ, Gardner E (1951) Prenatal development of the human elbow joint. Am J Anat 88:429–470
2. Loehr J, Uhthoff HK (1989) The pathomechanism of pulled elbow: a first review in the human embryo and fetus. (submitted for publication)

Chapter 10

The Development of the Wrist and Hand

K.A. CAUGHELL, A.H. MARTIN and H.K. UHTHOFF

Chondrification of wrist and hand progresses in a proximodistal direction. At stage 17, when the embryo measures approximately 12 mm, radius and ulna start to chondrify; both bones are widely separated. Subsequently, both styloid processes chondrify; they are well visible at stage 19. Also at this stage, elements of the carpus chondrify, starting with the capitate and hamate. The lunate and pisiform are the last to chondrify. By 6 weeks (Fig. 10.2) when the embryo measures approximately 18–20 mm, all bones but the distal phalanges can be identified. Of special interest is the formation of the os centrale situated between the scaphoid and capitate (Fig. 10.3). It is well seen toward the end of the embryonic period. During the fetal period it will fuse with the scaphoid. Before the fetal period starts, that is to say, at the end of the embryonic period, all elements of the hand have shapes, arrangement, and relationships we are used to seeing in adults (Figs. 10.4 and 10.7). At that time, sesamoids start to chondrify.

At stage 22 the formation of a bone collar around the radius and ulna is well established. It starts at stage 18 (see Chap. 9, Fig. 9.2). Vascular invasion follows at the beginning of the fetal period. At that time a bone collar forms around the second metacarpus [1]. Nobak and Robertson [2] observed this event a little later when the fetus measures 37 mm. The first metacarpus is the latest of the metacarpi to show evidence of ossification. Vascular invasion of carpal bones starts with the lunate at a time when the fetus measures 45 mm. Scanning electron microscopy shows particularly well the fibrous and chondrogenic layers of the perichondrium (Fig. 10.13). The growth of tubular bones is described in detail in Chap. 2.

Of particular interest is the ossification of the distal phalanges. Their chondrification is complete when the embryo measures 26 mm. Toward the end of the embryonic period (stage 23) a membranous ossification starts at the tips of phalanges II–V and is always present at the beginning of the fetal period. Intramembranous ossification of the tip of the thumb occurs when the fetus measures around 43 mm. When viewed on frontal sections the bone formation takes the form of a mushroom overhanging the cartilaginous part (Fig. 10.6). On sagittal sections the bony tip is only marginally thicker than the cartilaginous part (Fig. 10.8). This direct bone formation is followed by enchondral ossification which starts at the junction of the bony tip and the cartilage; obviously it can only progress proximally (Fig. 10.14a and 10.16a).

Formation of a bone collar of the proximal phalanges occurs when the fetus measures between 43 and 50 mm, the latest being the fifth digit. In the middle phalanx this activity starts in fetuses measuring between 56 and 70 mm and is well seen at 77 mm (Fig. 10.11). Vascular invasion of the proximal phalanges starts in the third digit when the fetus measures 65 mm and occurs in the first and fifth digit when it measures 83 mm. Advanced vascular invasion and enchondral ossification is seen in Fig. 10.14a. Vascular invasion of the middle phalanges is seen when the fetus measures 85 mm. By 16 weeks enchondral ossification of the distal radius has reached the metaphyseal level (Fig. 10.15). The progression of enchondral ossification (phase 1) distally and proximally can be well seen in the metacarpus (Fig. 10.15).

Early joint formation is evidenced by a three-layered interzone at the level of the wrist and some intercarpal joints. It can already be observed when the embryo measures 28 mm (stage 23; Fig. 10.3). With an increase in length of only 2 mm the same event occurs in all future finger joints. Cavitation implies the occurrence of cleft-like spaces in the middle, loose-celled layer of the interzone; it can be observed to start either at the periphery or centrally. It precedes by many weeks the presence of an actual joint cavity, lined with synovial tissue. This stage seems to be reached in all joints of wrist and hand when the fetus measures 115 mm (Fig. 10.14b). A joint cavity may not always be visible on histologic sections (Fig. 10.14a). This should not be interpreted as an absence of a joint, but rather as an artifact occurring during processing.

Collateral ligaments at the level of the metacarpophalangeal joints can be observed surprisingly early (stage 18). Of course at that time they are characterized by a cellular condensation rather than the presence of collagen fibers (Fig. 10.2). The latter form at the end of the embryonic period. Of particular interest is the formation of the transverse carpal ligament since it covers the carpal canal. It is well formed at stage 22. At this stage the canal contains easily identifiable superficial flexor tendons, the median nerve, and the sheet-like formation of deep flexor tendons (Fig. 10.1). Ulnar nerve and artery are clearly outside the canal. The contents of the carpal canal as well as its limiting structures are well seen at 8½ weeks (Fig. 10.5) and more so at later stages (Figs. 10.9b, 10.12, and 10.16b).

Cellular condensations indicating the presence of tendons can be recognized when the embryo measures 25 mm. Both flexor digitorum superficialis and profundus can be seen as separate entities at 6 weeks (Fig. 10.1). The formation of tendon sheaths occurs much later; it seems to coincide with the formation of joint cavities (Figs. 10.9a and 10.10). Although formation of muscles occurs earlier, they can be easily observed at 7 weeks (Fig. 10.3). Scanning electron microscopy does permit recognition of microtubules at 12 weeks (Fig. 10.13).

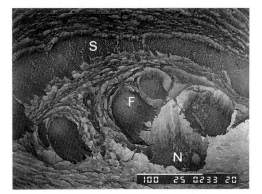

Fig. 10.1. Embryo L5, stage 17, 6 weeks, 13 mm. Scanning electron microscopy (SEM) of a cross section at the level of the carpal canal, showing the median nerve (*N*), the individual superficial flexors (*F*), and the sheet (*S*) of deep flexor tendons. The tendons are surrounded by a loose cellular stroma. *bar,* 100 μm

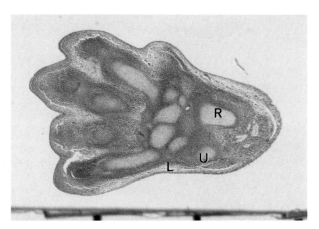

Fig. 10.2. Embryo 118N, stage 18, 6½ weeks, 15 mm. This section shows the early stage of chondrification. Note the distance between radius (*R*) and ulna (*U*), and the uniform interzones between the carpal elements, except for an indication of three layers between radius and scaphoid. The scaphoid is in a precartilaginous stage. Cellular condensations lateral and medial to the fifth metacarpophalangeal interzone constitute the formation of collateral ligaments (*L*). Toluidine blue, scale in millimeters

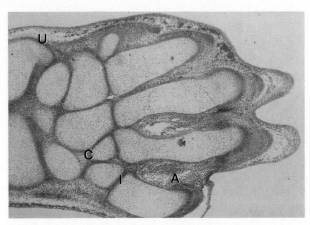

Fig. 10.3. Embryo 365L, 8 weeks, stage 23, 28 mm. In a little more than 1 week the cartilaginous structures have developed very fast. Both forearm bones are now closer together and the styloid process of the ulna (*U*) is present. Note the less mature chondrification of the os centrale (*C*) distal to the scaphoid. This separate anlage will later fuse with the scaphoid. The interzone (*I*) between trapezium and base of the first metacarpus is still flat and not saddle-shaped. Abductor indicis (*A*). H & E, same magnification as Fig. 10.2

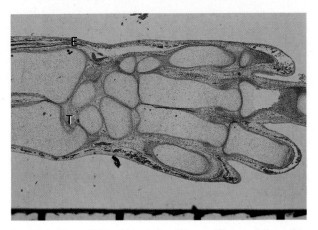

Fig. 10.4. Fetus 270R, 8½ weeks, 37 mm. The modeling of long bones is not complete. The triangular cartilage (*T*) distal to the ulna can be identified by a condensation of cells. Lateral to the radial styloid the extensor pollicis brevis (*E*) and the abductor longus tendon can be seen. All of the joint interzones are now three-layered. The collateral ligaments are well developed. HPS, scale in millimeters

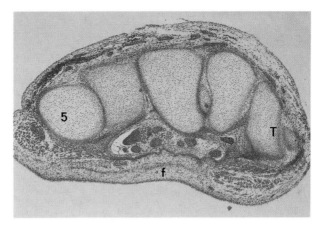

Fig. 10.5. Fetus 309, 8½ weeks, 34 mm. This section at the carpometacarpal level permits one to recognize the contents of the carpal canal. It is bordered by trapezium (*T*), trapezoid, capitate, hamate, and base of the fifth metacarpus (5). Lateral to the trapezium the base of the first metacarpus can be seen. Tendon sheaths have not completely formed yet. The transverse carpal ligament borders the canal at its palmar side. The *blue line* in the overlying tissue represents the palmar fascia (*f*). Muscles of the thenar and hypothenar eminences are well developed. H & E, same magnification as Fig. 10.2

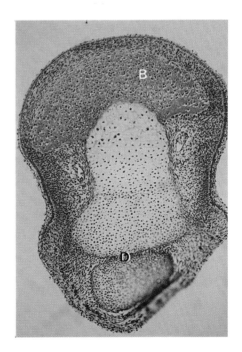

Fig. 10.6. Fetus 369R, 9 weeks, 45 mm. This section through the distal phalanx of the middle finger illustrates well the process of ossification which is different from that of the proximal and middle phalanges. Note the mushroom-like lateral overhang of intramembranous bone formation (*B*). Distal interphalangeal joint (*D*). Azan, ×75

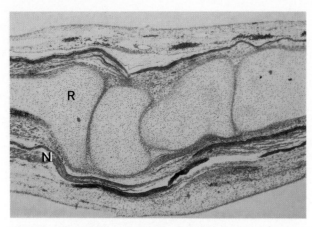

Fig. 10.7. Fetus 351L, 9½ weeks, 50 mm. Section of the wrist, showing radius (*R*), lunate, capitate, and third metacarpus. The median nerve (*N*) and the palmaris longus as well as part of the common extensor tendons are seen. The pronator quadratus can be identified. The three layers of the interzones are visible. H & E, same magnification as Fig. 10.2

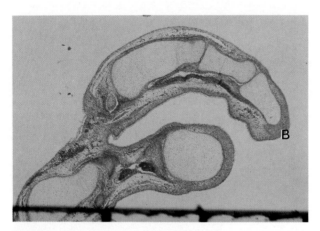

Fig. 10.8. Fetus 373R, 10 weeks, 56 mm. Compare this sagittal section with Fig. 6. In this plane the bone formation (*B*) at the tip of the thumb is not larger than the cartilage model. The nail field has not developed yet. Note the long flexor tendon of the thumb. The sesamoid at the metacarpophalangeal joint starts to form. The index finger is seen in cross section. Azan, scale in millimeters

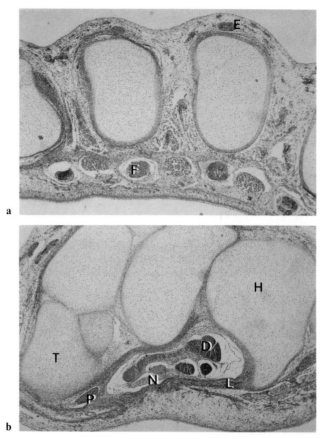

Fig. 10.9. a Fetus 377, 11 weeks, 70 mm. This section at the metacarpal level shows the superficial and deep flexor tendons (*F*) in their sheath; in between, the neurovascular structures and the lumbricales muscles. The extensor tendons are easily recognizable (*E*). Azan, same magnification as in Fig. 10.2. **b** Same fetus. This section goes through the carpal canal at the level of the hamate (*H*) and triquetrum (*T*). Whereas the superficial flexors can be recognized as isolated structures in their tendon sheaths, the deep flexors (*D*) only start to separate. Note also the median nerve (*N*). The flexor pollicis longus (*P*) is clearly outside the canal, as is the radial artery. Note the strong transverse ligament (*L*) which is better visualized on the ulnar side. Azan, same magnification as Fig. 10.2

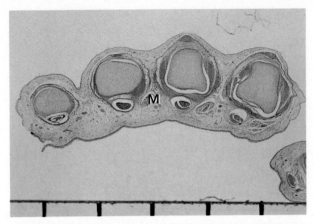

Fig. 10.10. Fetus 194, 12 weeks, 75 mm. Section at level of metacarpal heads showing their intra-articular position and the strong capsule. Superficial and deep flexor tendons are clearly separated. The extensor hood is well developed. Nerves, vessels, and lumbrical muscles (*M*) can be recognized in their compartments. Toluidine blue, scale in millimeters

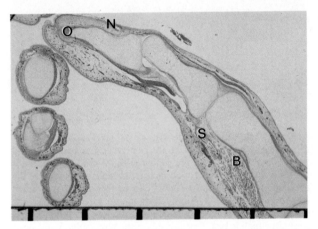

Fig. 10.11. Fetus 370L, 12 weeks, 77 mm. Section of a thumb showing the membranous ossification (*O*) of the distal phalanx. No cellular and vascular invasion has taken place in either of the two phalanges. The sesamoid of the metacarpophalangeal (*S*) joint is well formed as is the sesamoid of the interphalangeal joint. The latter is clearly inside the joint capsule, but not in the flexor tendon. The nail field (*N*) has developed. Three other fingers are seen in cross section. Periosteal bony sleeve of first metacarpus (*B*). Azan, scale in millimeters

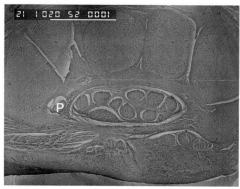

Fig. 10.12. Fetus 209L, 12 weeks, 70 mm. SEM cross section at the level of the carpal canal. The palmaris longus tendon is situated superficial to the transverse carpal ligament. The contents of the carpal canal, consisting of median nerve and flexor tendons, are easily identifiable. Joint cavities exist between the carpal bones. Flexor pollicis longus (*P*). Scale in micrometers

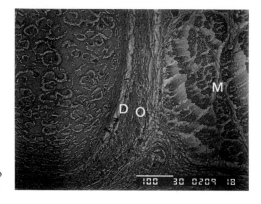

Fig. 10.13. Fetus 149N, 12 weeks, 76 mm. SEM cross section illustrating both layers of the perichondrium, the outer layer (*O*) is fibrous and deep to it the chondrogenic layer (*D*). The perichondrium surrounds the hyaline cartilage. Its chondrocytes are contained in lacunae whereas cell nests are located more centrally. Note the microtubules (*M*) of the developing muscle adjacent to the perichondrium. Scale in micrometers

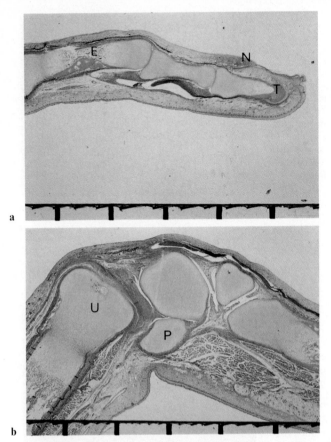

Fig. 10.14. a Fetus 173R, 14 weeks, 115 mm. Section of a finger. Vascular invasion of the bony tuft (*T*) of the distal phalanx; it progresses proximally toward the bone-cartilage junction where enchondral ossification will start. Deep and superficial flexor tendons are seen. Enchondral ossification (*E*) of the proximal phalanx progresses distally and proximally. The fingernail (*N*) is forming. Azan, scale in millimeters. **b** Same embryo. This section goes through ulna (*U*), triquetrum, hamate, and fifth metacarpus. The triangular cartilage is well seen as is the pisiform bone (*P*). Joint cavities are now lined by synovial tissue. Azan, scale in millimeters

Fig. 10.16. a Fetus 341L, 18 weeks, 158 mm. Section illustrating well the particular process of ossification of the distal phalanx where enchondral ossification (*O*) is now more important than the membranous one. The deep flexor tendon (*T*) inserts into the fibrous perichondrium. Nail (*N*) and nail bed are well formed. Tactile corpuscles (Golgi-Mazzoni) are seen in the fingertips besides many sweat glands. Azan, scale in millimeters. **b** Same fetus. Cross section through the carpal canal. The transverse ligament spans from trapezium (*T*) to hamate (*H*). Palmar and radial to it, the muscles of the thenar eminence. The flexor carpi radialis (*R*) is outside the carpal canal. Note median nerve (*N*) and flexor tendons. Cartilage canals are seen in the carpal bones. Azan, scale in millimeters

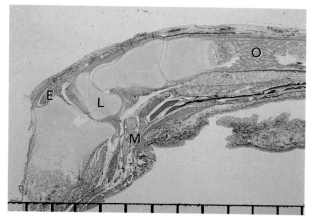

Fig. 10.15. Fetus 344R, 16 weeks, 135 mm. Section through radius, lunate (*L*), capitate, and third metacarpus. Cartilage canals can be seen in the radial epiphysis, lunate, and capitate. The median nerve (*M*), flexor and extensor tendons, intrinsic muscles as well as the extensor pollicis longus (*E*) are visible. Note the enchondral ossification (*O*) of the metacarpus. Azan, scale in millimeters

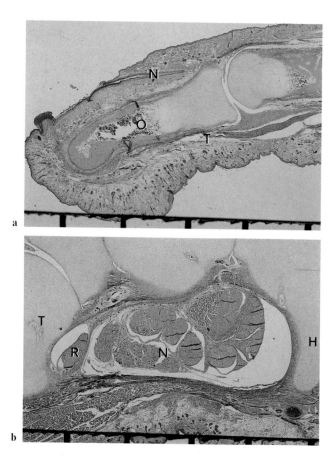

References

1. Gray DJ, Gardner E, O'Rahilly R (1957) The prenatal development of the skeleton and joints of the human hand. Am J Anat 101:169–223
2. Noback CR, Robertson GG (1951) Sequences of appearance of ossification centers in the human skeleton during the first five prenatal months. Am J Anat 39:1–28
3. O'Rahilly R, Gardner E (1972) The initial appearance of ossification in staged human embryos. Am J Anat 134:291–308

Chapter 11

The Development of the Pelvis

J.P. McAuley and H.K. Uhthoff

Each hemipelvis develops from one cartilaginous anlage. According to Adair [1] cartilage formation inside the blastemic condensation of the pelvis starts at more than one site. Our material does not support this observation. At 8 weeks a single cartilaginous anlage is well seen (Fig. 11.3), with no transitional zone between the ilium, the ischium, and the pubis. This is evident in sections through the acetabulum (Fig. 11.7 b). In our material, evidence of ossification is first observed in the ilium where a periosteal direct bone formation is seen at 9½ weeks (Fig. 11.4). Shortly thereafter the area of degenerating chondrocytes (Streeter phase 5) in the middle of the ilium is the site of cellular and vascular invasion. This marks the beginning of the first phase of enchondral bone formation, a process well known in long bones (see Chap. 2). Gardner [5] also observed the beginning of ossification of the ilium at 9 weeks. The anterior superior iliac spine is said to develop at the end of the 3rd month [8]. The pubis starts to ossify around the 12th week and the ischium around the 15th week [5]. Ossification of ilium and ischium is seen in Chap. 12, Fig. 12.15.

Particularly at the pelvis, the description and the dates of appearance of ossification lack precision and can be contradictory. First, authors fail to state what is meant by bone formation. Is it periosteal or is it inside the cartilage anlage? Secondly, some authors base the presence of ossification on microscopic examinations and others on radiologic findings.

At the level of the symphysis pubis, the hemipelves are separated by a narrow band of rather densely packed cells. This is already obvious at stage 23 according to Gasser [7]. Figure 11.5 shows the symphysis at 9½ weeks. Vascular invasion of the pubic cartilages very close to the symphysis starts from the inferior aspect and is particularly well seen at 18 weeks (Fig. 11.11). Most of the stability of the symphysis is provided by the inferior pubic ligament [4] (Fig. 11.11).

Of particular interest is the development of the sacroiliac joint. The future joint is already recognizable at 6 weeks (Fig. 11.1) when the interzone consists of a cell layer of uniform density. As early as 7 weeks all three layers of the interzone become visible (Fig. 11.2). From there on the development of the sacroiliac joint is quite different from that of other joints. The first characteristic of this joint is the simultaneous presence of a synovial joint at the caudal part and a synarthrosis at the middle and cranial parts. A clear difference in development between the caudal part of the sacroiliac joint and the middle and cranial parts is obvious at 14 weeks (Fig. 11.9). Cavitation at the caudal part leads to the formation of a joint space. More cranially, richly vascularized tissue occupies the space between sacrum and ilium. This distinction becomes more striking at 16 weeks (Fig. 11.10) with loose connective tissue present between the future bones in the middle part and a joint space caudally. Formation of dense ligamentous structures can be observed at the upper part at 10 weeks (Fig. 11.6) and at 16 weeks (Fig. 11.14).

The second characteristic is the different temporal sequence in the development of each side of the joint, the development of the ilium being ahead of that of the sacrum. The two components of the sacroiliac joint, the ilium and the sacrum, are already obvious at 6 weeks (stage 18; Fig. 11.1). At this stage the chondrocytes of the ilium are in Streeter phase 3 whereas those of the adjoining part of the sacrum are in Streeter phase 2. During the following week the chondrocytes of the ilium progress to Streeter phase 4 and those of the sacrum to Streeter phase 3. The difference in development of both bones in time persists throughout the period of observation (Figs. 11.7a, 11.8, 11.9, 11.10, 11.12, and 11.13).

The difference in development between the sacrum and ilium is not limited to the temporal development of the cartilaginous anlagen; it is also seen at the level of the joint. Whereas all known diarthrodial joints are covered by hyaline cartilage on both surfaces, in the sacroiliac joint the iliac surface consists of fibrocartilage and the sacral side of hyaline cartilage [2]. This is well documented in Fig. 11.12, taken from a 20½-week-old fetus.

Direct bone formation in the caudal part of the ilium is evident by 9½ weeks. Enchondral bone formation phase 1 is also seen. At that time chondrocytes of the sacrum have advanced to phase 4 (Fig. 11.4). Whereas at the cervical, thoracic, and lumbar levels the nerve roots exit laterally through the intervertebral foramina, at the sacrum they exit anteriorly (see Chap. 7, Fig. 7.9).

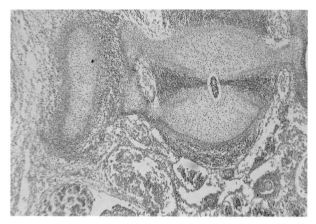

Fig. 11.1. Embryo 86N, 6 weeks, stage 18, 13 mm. Frontal section. Sacral vertebrae 1 and 2 separated by an intervertebral disk and containing part of the notochord can be seen. The ala of the sacrum is separated from the ilium by cells of moderate but uniform density. Chondrocytes of the ilium are in Streeter phase 3 and those of the sacral ala in Streeter phase 2. Goldner, × 40

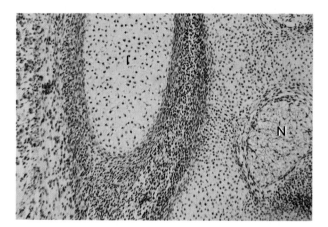

Fig. 11.2. Embryo 119N, 7 weeks, stage 19, 21 mm. Frontal section. A higher magnification of the lower pole of the sacroiliac joint clarifies the three layers of the interzone. Note the nerve root (*N*) in the sacral foramen. Obviously the chondrocyte maturation is more advanced in the ilium (*I*) than in the sacrum. HPS, × 100

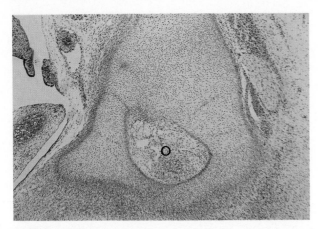

Fig. 11.3. Embryo 72, 8 weeks, stage 22, 27 mm. Frontal section. This photomicrograph shows well that the hemipelvis is one cartilaginous anlage. Obturator foramen (*O*) and nerve are seen. HPS, ×40

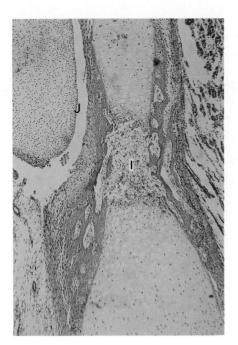

Fig. 11.4. Fetus 198N, 9½ weeks, 48 mm. Frontal section. This section demonstrates the rather extensive periosteal sleeve and the cellular invasion of the ilium (*I*). The joint cavity (*J*) at the caudal part of the sacroiliac joint is obvious. More cranially, however, no similar cavity is seen. (The cleft surrounding the lower pole of the sacrum is an artifact). Goldner, ×40

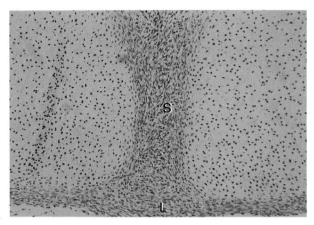

Fig. 11.5. Same fetus as in Fig. 11.4. Frontal section. The symphysis pubis (*S*) is seen as a blastemic condensation between the anlagen of the hemipelves. The cellular condensation at its inferior part shows early formation of the inferior pubic ligament (*L*). HPS, × 100

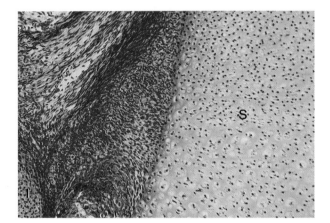

Fig. 11.6. Fetus E6, 10 weeks, 50 mm. This transverse section made at the cranial part of the sacrum (*S*) shows the strong sacroiliac ligament. Ilium (*I*). Azan, × 100

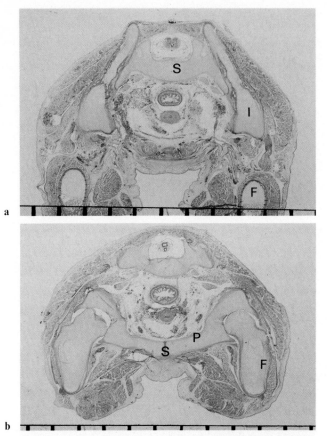

Fig. 11.7 a, b. Fetus 379, 10½ weeks, 60 mm. Transverse section. **a** This section shows advanced periosteal bone formation of both iliac bones (*I*), the cartilaginous sacrum (*S*) (as one anlage, not yet fused posteriorly) and both femora (*F*). Azan, scale in millimeters. **b** This section from the same fetus is more caudally situated. It shows that the superior pubic ramus (*P*) and the ilium are one cartilaginous anlage. Note also the symphysis pubis (*S*). Its syndesmosis is very narrow. Both proximal femora (*F*) are well shown. Azan, scale in millimeters

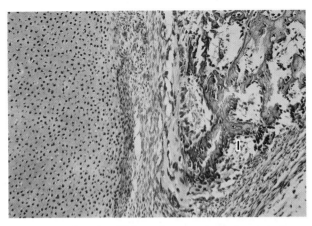

Fig. 11.8. Fetus 48, 12 weeks, 75 mm. Frontal section. This section passes anterior to the lower part of the sacroiliac joint. It shows advanced bone formation of the ilium (*I*). The joint cavity does not reach into this area. Azan, × 100

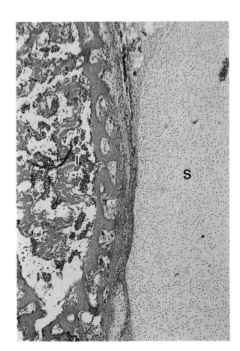

Fig. 11.9. Fetus 23N, 14 weeks, 105 mm. Frontal section. In this section through the sacroiliac joint between lower and middle thirds, the bone formation of the ilium (*I*) has already progressed in a cranial direction. Hematopoietic elements have formed. There is further progression in chondrocyte maturation in the sacral ala (*S*). The difference in development between both sides of the joint is impressive. The joint space itself is not visible. Goldner, × 40

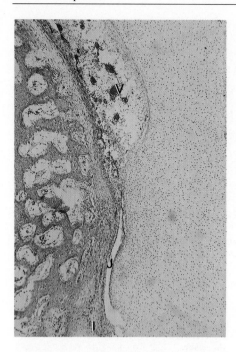

Fig. 11.10. Fetus 24N, 16 weeks, 130 mm. Frontal section. This section demonstrates well the formation of a synovial joint (*J*) at the caudal part and that of a synarthrosis further cranially. Cartilage islands of the ilium (*I*) persist next to the joint caudally. The rich vascularity (*V*) of the middle part of the sacroiliac joint is striking. Azan, × 40

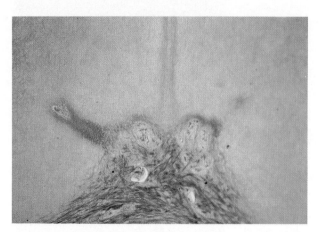

Fig. 11.11. Fetus 1186-19, 18 weeks, Carnegie Collection. Frontal section. The vascular invasion of both pubic rami close to the symphysis pubis starts caudally. The synchondrosis is well seen as is the strong inferior pubic ligament. Azan, × 16

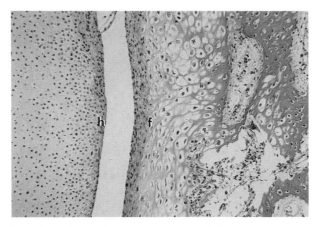

Fig. 11.12. Fetus 146, 20½ weeks, 195 mm. Frontal section. The synovial joint at the lower part of the sacroiliac joint is well formed. Hyaline cartilage (*h*) borders it at the sacral side whereas on the iliac side a thin layer of fibrocartilage (*f*) is supported by bone. Goldner, ×100

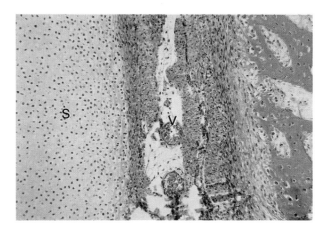

Fig. 11.13. Same fetus as in Fig. 11.12. Frontal section. The middle part of the joint is a syndesmosis mainly composed of fibrous and loose connective, richly vascularized tissue. Note that no bone formation has yet occurred in the sacral ala. Goldner, ×100

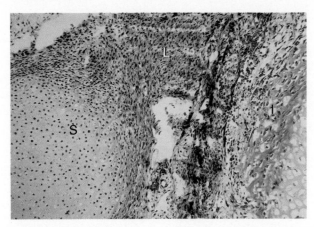

Fig. 11.14. Fetus 24N, 16 weeks, 130 mm. Frontal section. At the upper part strong ligaments (*L*) can be seen uniting the sacrum (*S*) to the iliac bone (*I*). Note the distance between ilium and sacrum which is wider than in the more caudal parts. Goldner, ×100

References

1. Adair FL (1918) The ossification centers of the fetal pelvis. Am J Obstet 78:175–199
2. Bowen V, Cassidy JD (1981) Macroscopic and microscopic anatomy of the sacroiliac joints from embryonic life and the eight decade. Spine 6:620
3. Frances CL (1951) Appearance of centers of ossification in the human pelvis before birth. AJR 65:778–783
4. Gamble JG, Simmons SC, Freedman M (1986) The symphysis pubis. Clin Orthop 203:261–272
5. Gardner E (1971) Osteogenesis in the human embryo and fetus. In: Bourne GH (ed) The biochemistry and physiology of bone, vol 33, 2nd edn. Academic, New York, pp 77–118
6. Gardner E, Gray DJ (1950) Prenatal development of the hip joint. Am J Anat 87:163–211
7. Gasser RF (1975) Atlas of human embryos. Harper and Row, Hagerstown
8. Lambertz J (1900) Die Entwicklung des menschlichen Knochengerüstes während des fötalen Lebens. Fortschr Röntgenstr [Ergänzungsh] 1
9. Schunke GB (1938) The anatomy and development of the sacroiliac joint in man. Anat Res 72:3

Chapter 12

The Development of the Hip

H.K. UHTHOFF and T. CAREY

The position of flexion of the coxofemoral joint during the intrauterine period often makes it difficult to interpret sections crossing the joint in various planes. From our studies of anatomy and from our clinical expsosure we are used to see this joint in extension. We will, however, attempt to ease the visualization of the coxofemoral joint in a flexed position by giving as dynamic as possible a picture of the successive stages of formation of this joint between the 6th and 20 weeks.

As customary, the direction of planes used in our descriptions refers to the trunk. Consequently, because of hip flexion, frontal sections will show the femoral head as a sphere, sometimes with neck and greater trochanter, and, if the sections cross the femoral head at its center, the superior and inferior labrum will be seen in a slightly oblique direction owing to the obliquity of the acetabulum. Sagittal sections permit one to see part of the femur or its entirety, depending on the degree of abduction or adduction; the anterior and posterior labrum will be crossed in a slightly oblique direction, a fact which precludes an assessment of the coverage of the femoral head. Depending on the level, some sagittal sections show the more cranial part of the anterior labrum and the more caudal part of the posterior labrum. Finally, transverse sections will show more or less of the femur, depending on the degree of flexion; they will demonstrate the state of the anterior and posterior labrum, as well as the degree of acetabular anteversion.

The earliest evidence of the hip joint can be observed when the embryo is 6 weeks of age and barely measures 1.5 cm. An interzone of blastemic condensation separates the cartilaginous anlage of the femur from that of the pelvis (Fig. 12.1). The cells of this zone have round nuclei.

A few days later the uniform interzone undergoes a loosening in its midportion where concomitantly the round cells of the middle layer of the interzone become elongated. The outer layers of the interzones will form the chondrogenous layers (Fig. 12.2). They are in continuity with the perichondrium. At 7 weeks the coxofemoral joint becomes recognizable, even to the unskilled eye (Fig. 12.3). When the embryo measures less than 3 cm in length we can speak of a joint. At 8 weeks cavitation is complete although it is limited in its extent (Figs. 12.4 and 12.6).

The labrum is well formed at 8 weeks and an impressive growth, especially of the iliac part of the pelvis, has occurred (Fig. 12.3). Later on, as seen at 12, 16½ and 18 weeks, the superior and inferior labrum stand out clearly as distinct structures (Figs. 12.9, 12.14 and 12.15). Of particular interest is the fact that the capsule does not insert into the labrum, but blends with the fibrous layer of the perichondria (Figs. 12.10 and 12.11). The synovial lining with its subsynovial tissue, however, forms a recessus separating the labrum from the capsule (Figs. 12.11 and 12.13).

Directing our attention to the growth of the acetabulum and the femoral head, we observe that hypertrophy of the cartilage cells of the pelvis takes place at 7 weeks and is ahead of a similar development of the femoral head. By 8 weeks the maturation of chondrocytes of pelvis and femoral head have reached the same phase (Fig. 12.4). Vascular invasion of the femoral epiphysis which starts at the site of the synovial reflections can be easily observed at 10 weeks (Fig. 12.8). At a little more than 3 months of age the progression of enchondral ossification of the pelvic bones, along with a concomitant increase in thickness of the periosteal bony sleeve, is the most impressive feature (Fig. 12.11).

Muscle and joint capsule can be easily identified at 8 weeks at which time the greater trochanter is appearing (Fig. 12.6). It is fascinating to see in such a small embryo all elements we find in adults, although their shape and relative size will change with further growth. Of particular interest is the extent of the cartilaginous coverage of the femoral head at the posterior aspect (Fig. 12.16). In one specimen this has permitted a posterior subluxation (Fig. 12.12). Whereas at 10 weeks the diameters of the head and neck are almost equal (Figs. 12.7 and 12.8), at 16 weeks the effect of modeling of the neck is evident: the ratio between head and neck approaches values we find in the postnatal period (Fig. 12.13).

At 18 weeks the joint cavity has attained its final shape. The synovial lining with its subsynovial fatty tissue is clearly identifiable. The gluteus medius, which inserted into the fibrous layers of the perichondrium of the femur distal to the greater trochanter in earlier periods, attaches to the greater trochanter directly; Sharpey's fibers can be identified at this site. This is well seen at full term (Fig. 12.17).

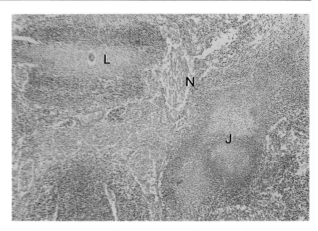

Fig. 12.1. Embryo 117N, stage 18, 6 weeks, 14 mm. Frontal section. The cartilaginous anlagen of pelvis and femoral head are separated by a blastemic condensation of uniform cellular density (*J*). The hip joint is at the level of a lumbar vertebra (*L*) which contains part of the notochord. Between lumbar spine and pelvis the lumbar plexus (*N*) can be seen. Reconstruction of serial sections showing the third dimension documents that all three bones of the hemipelvis form one cartilaginous anlage. Goldner, × 40

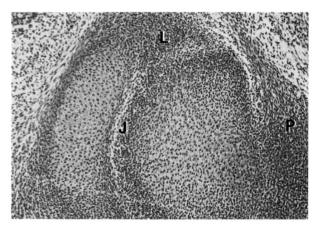

Fig. 12.2. Embryo 47N, stage 18, 6½ weeks, 15 mm. Sagittal section. The three layers of the interzone are seen (*J*). The chondrogenous zone of the acetabulum ends superiorly and inferiorly in a cellular condensation which is interpreted as being an early labrum (*L*). The dense tissue in front of the femoral head represents the psoas muscle (*P*). Azan, × 100

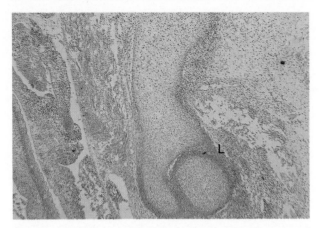

Fig. 12.3. Embryo 123, stage 19, 7 weeks, 19 mm. Frontal section. Early cavitation is seen with strands of cells traversing the cavity. Cellular condensations indicating the formation of the capsule and the superior labrum (*L*) are distinct. Note the cartilage maturation of the pelvis (Streeter phase 4) which is ahead of that of the femoral head (Streeter phase 3). Goldner, × 40

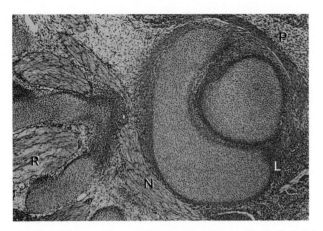

Fig. 12.4. Embryo 72, stage 22, 8 weeks, 27 mm. Sagittal slightly oblique section. The section passes through the femoral head medial to the shaft; therefore, the head is seen as a sphere. Both labra (*L*) can be recognized by cellular condensations. The iliopsoas (*P*) is well formed; note how big the sciatic nerve (*N*) is, particularly when compared with the size of the other elements. The obliquity of the section is evidenced by the presence of nerve roots (*R*) exiting between pedicles. HPS, × 40

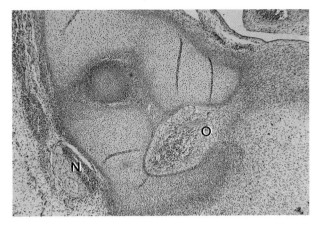

Fig. 12.5. Same embryo as in Fig. 12.4. Sagittal but slightly oblique section. This section of the opposite hip goes partly through the obturator foramen which contains the obturator nerve (*O*). The femoral head lies in the acetabular cavity. The sciatic nerve (*N*) is cut tangentially. Goldner, × 40

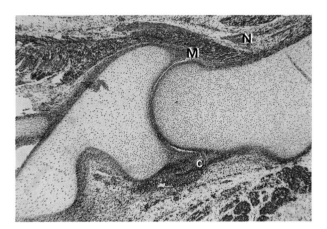

Fig. 12.6. Embryo 151N, stage 23, 8 weeks, 27 mm. Sagittal section. This section illustrates the flexed position of the hip. The joint cavitation, the superior and inferior labrum, the joint capsule (*c*), the iliopsoas muscle (*M*), and the femoral nerve (*N*) are well seen as is the greater trochanter. The femoral neck is nearly as wide as the head. This section crosses the joint more laterally than that in Fig. 12.4. Goldner, × 40

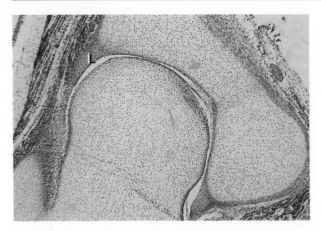

Fig. 12.7. Fetus 53N, 9½ weeks, 45 mm. Transverse section. This transverse cut goes through the anterior and posterior (*L*) labrum, the acetabular fossa, and the fovea capitis. Some strands of the ligamentum capitis femoris are seen. The gluteus medius and maximus are easily recognizable. It is well worth noting the lack of posterior coverage of the head by the acetabular cartilage, especially when compared with the anterior coverage. Note that the capsule is not attached to the anterior labrum. In Chap. 11, Figure 11.6 is taken from an embryo of the same age and shows ossification of the ilium. HPS, × 40

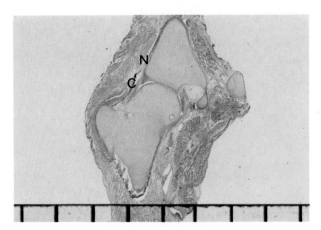

Fig. 12.8. Fetus 112N, 10 weeks, 53 mm. Transverse section. The interpretation of this transverse section is rendered difficult by a certain obliquity of the section and the flexed position of the hip. The ligamentum capitis femoris in the acetabular fossa and the anterior and posterior labrum are well seen. Whereas the typical cellular condensation identifying the capsule cannot be seen in Fig. 12.7, in this section the posterior capsule (*C*) is well defined. The acetabular fossa is well seen. Note the vascular invasion of the femoral head which, as in all joints, occurs at the sites of synovial reflections. The sciatic nerve (*N*) is cut obliquely. As in Fig. 12.7, head and neck have an identical diameter. Goldner, scale in millimeters

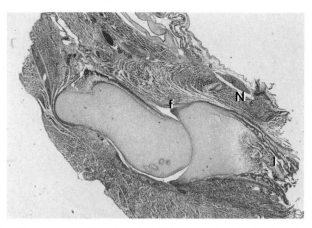

Fig. 12.9. Fetus 109N, 12 weeks, 78 mm. Sagittal section. A fold between the inferior capsule and labrum is seen and a similar process can be seen superiorly (*f*). It represents a synovial reflection and shows that the capsule does not insert into the labrum. The joint cavity englobes the entire head. Enchondral bone formation and periosteal ossification of the ilium (*I*) are seen. Anteriorly, the femoral nerve (*N*) is well seen. Note that head and neck are still of the same diameter. HPS, same magnification as in Fig. 12.8

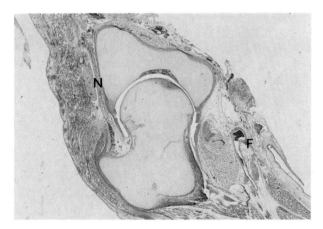

Fig. 12.10. Fetus 26N, 12 weeks, 80 mm. Transverse section. slightly oblique from anterolateral to posteromedial. The distinction between joint capsule and labrum is becoming clearer than at earlier stages. The neck shows evidence of modeling, effectively reducing its diameter. The acetabular fossa and the fovea capitis are well seen, as is the obliquely cut sciatic nerve (*N*). The femoral nerve and artery (*F*) are separated from the femur by the rectus femoris muscle. Azan, same magnification as in Fig. 12.8

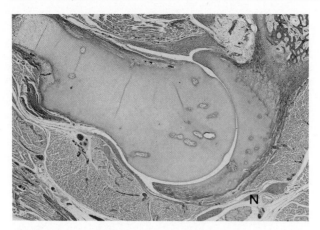

Fig. 12.11. Fetus 105N, 14½ weeks, 112 mm. Sagittal section. This section shows well the advancing ossification in the iliac bone. Overlying the anterior capsule the reflected head of the rectus femoris muscle can be seen. The sciatic nerve (*N*) lies behind the short rotator muscles. Both labra are free standing. Azan, same magnification as in Fig. 12.8

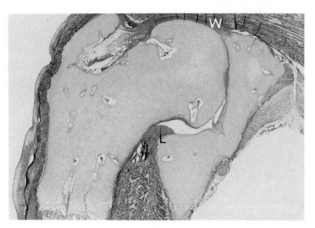

Fig. 12.12. Fetus 82N, 14 weeks, 105 mm. Transverse section. This illustrates a fetal hip with an insufficient posterior wall (*W*), the head is subluxed in a posterior direction, and the anterior labrum (*L*) is folded in. Note the multitude of vascular canals in femur and pelvis. Goldner, same magnification as in Fig. 12.8

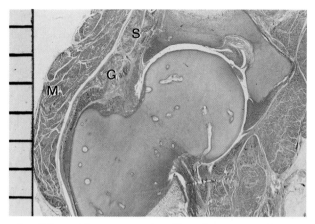

Fig. 12.13. Fetus 104N, 15½ weeks, 128 mm. Transverse section. Note the strong and increasingly more fibrous labrum, anteriorly as well as posteriorly, the acetabular fossa, and the ligamentum capitis femoris. The sciatic nerve is well defined. The gluteus medius (*G*) inserts into the perichondrium of the femur; it is covered by the gluteus maximus (*M*). Note the incomplete cartilaginous coverage of the femoral head posteriorly. The femoral neck is now smaller than the head. Sciatic nerve (*S*). Azan, scale in millimeters

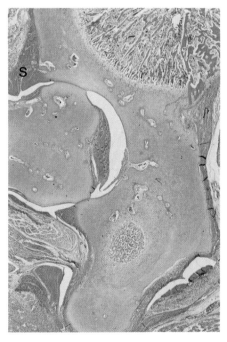

Fig. 12.14. Fetus 96N, 16½ weeks, 140 mm. Sagittal section. This section shows the ligamentum capitis femoris. The fossa acetabuli has been cut at its medial border. The superior capsule (*S*) does not insert into the labrum. Note the rich vascularization of the ligamentum capitis femoris and of the synovial membrane. Also of interest is the presence of cartilage canals close to the fovea capitis, confirming the role the ligamentum capitis femoris plays in the vascularization of the femoral head during development. Goldner, same magnification as in Fig. 12.13

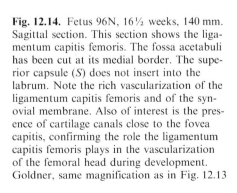

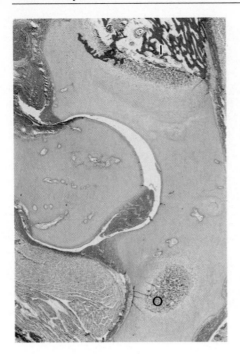

Fig. 12.15. Fetus 78N, 18 weeks, 160 mm. Sagittal section. This shows again the flexed position of the hip. The ossification of the ilium (*I*) is approaching the acetabulum. The ossific nucleus of the ischium (*O*) is less advanced. Both labra are seen, as is the ligamentum capitis femoris. Azan, same magnification as in Fig. 12.13

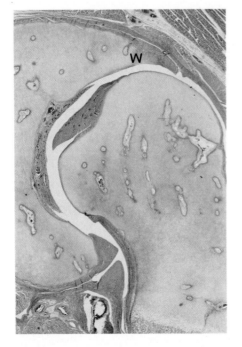

Fig. 12.16. Fetus 115N, 20 weeks, 185 mm. Transverse section. This illustrates the thrust of the flexed femur on the posterior acetabular wall (*W*). This wall consists of an elongated labrum, a rather thin capsule, and a cartilage which does not provide good coverage of the femoral head at its posterior aspect. Note again the rich vascularization of the acetabular fossa. Goldner, same magnification as in Fig. 12.13

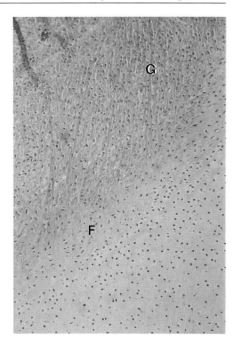

Fig. 12.17. Fetus 1556A, at full term, 320 mm. Transverse section. This microphotograph illustrates the insertion of the gluteus medius (*G*) into the greater trochanter. This is the first time Sharpey's fibers (*F*) inserting into the cartilage are seen at this level. At earlier stages all fibers of the muscle were continuous with those of the perichondrium of the femur. Some fibers of the muscle are still maintaining this relationship. H & E, × 100

Reference

1. Gardner E, Gray DJ (1950) Prenatal development of the human hip joint. Am J Anat 87:163–211

Chapter 13

The Development of the Knee

M.A. FINNEGAN and H.K. UHTHOFF

The knees of most specimens are in a position of flexion when studied. This position must be remembered when viewing the sections which were done in different planes in order to study all joint components. At 6 weeks, stage 17, femur, tibia, and fibula start to chondrify (Fig. 13.1); 1 week later (stage 19) the chondrification of the femoral condyles and tibia and fibula is well advanced (Fig. 13.2). Vascular canals are starting to invade the patella, femur, and tibia around the 12th week (Fig. 13.8). A tremendous progression is noted at 14 weeks. At this stage enchondral ossification of femur and tibia has reached the metaphyseal areas (Fig. 13.10 a, b). Formation of columns of chondrocytes is seen at 18 weeks (Fig. 13.14).

During intrauterine development the knee joint progresses from a homogeneous blastemic cell condensation to an articulation, complete with menisci, cruciate and collateral ligaments, and an extensor mechanism. A uniform interzone between femur and tibia is present at 6 weeks (Fig. 13.1). The formation of three layers starts at 6½ weeks and progression is seen at 7 weeks (Fig. 13.2). At 8 weeks joint cavitation starts, as seen at the patellofemoral level (Fig. 13.3). The cavitation of the femorotibial compartment begins first between femur and menisci at 9½ weeks (Fig. 13.6), and is seen between menisci and tibia shortly after (Fig. 13.7). With further development the joint cavity extends proximally under the quadriceps (Fig. 13.10 b).

The intra-articular structures develop in situ from the middle layer of the interzone. Whereas cellular condensations at the periphery of the joint point to an early formation of collateral ligaments at 7 weeks (Fig. 13.2) the cruciate ligaments become obvious at 8 weeks (Fig. 13.3). They are well developed at 9½ weeks (Fig. 13.4). Menisci can be identified at 9½ weeks (Fig. 13.6). With time their collagen content increases (Fig. 13.7). At 16 weeks blood vessels can be well recognized at the attachments of the menisci. In no specimen was fibrocartilage found; this is not surprising as fibrocartilage usually develops in fibrous tissues subjected to sustained pressure. Therefore, it is postulated that the presence of this tissue in menisci is related to weight bearing. In our material discoid menisci were not observed in any stage of development. We can therefore not subscribe to the commonly held view that all menisci go through a phase of discoid structure. A meniscus at nearly 20 weeks shows dense fibrous tissue and a richly vascularized rim (Fig. 13.15). This could explain the potential for repair of a peripheral tear as compared with a central one.

The formation of cruciate ligaments is obvious at 8 weeks (Fig. 13.3). Like all ligaments, they consist of cellular condensations at the beginning; collagen formation increases with time (Fig. 13.7, and 13.10 a, b). The presence of collateral ligaments is observed at 7 weeks (Fig. 13.2), their identification becomes easy at 10 weeks (Fig. 13.7), and more so at 16 weeks (Fig. 13.13).

Despite the fact that the patellofemoral joint develops rather early, both retinacula can only be recognized at 9½ weeks (Fig. 13.5). The lateral retinaculum appears as a dense fibrous structure. The medial retinaculum, on the other hand, consists of rather loosely arranged tissue (Fig. 13.16). The patella is not exactly centered in the femoral groove, but tends to ride more laterally. Both features could predispose to a lateral tracking of the patella, resulting in increased lateral pressure. Early on, the future infrapatellar fat pad is a distinct vascular area (Fig. 13.3, 13.4, and 13.8). Later on, rather loose cellular tissue occupies this area (Fig. 13.10 b) and formation of adipose cells is observed at 18 weeks. Strands of cells extending from the fat pad to the intercondylar notch are frequently observed [4]. Condensation of cells at future sites of muscles is seen at 7 weeks (Fig. 13.2), but identification of various muscles is only possible starting at 14 weeks (Fig. 13.10 a, b).

The formation of the anterior tibial tuberosity merits special attention [3, 5]. At 12 weeks a vascular channel invades the proximal tibial cartilage anlage in an upward and oblique direction and starts to separate a tongue-shaped cartilage from the epiphysis (Fig. 13.9); 2 weeks later the vessel has grown in deeper (Figs. 13.10 a and 13.11). By 18 weeks the formation of the tuberosity is better outlined (Fig. 13.14); it is well defined at 20½ weeks (Fig. 13.17). When the fetus has reached the 20th week of gestation, all anatomic structures found during the postnatal period are easily recognizable.

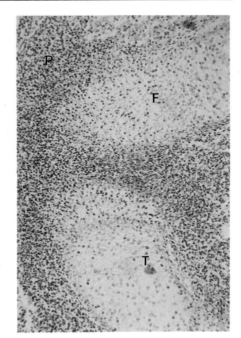

Fig. 13.1. Embryo 117N, stage 17, 6 weeks, 14 mm. Sagittal section. The chondrifying femur (*F*) and tibia (*T*) are at right-angles to each other. The interzone is wide, but of uniform cellular density. Anterior to the lower end of the femur a blastemic condensation is seen, representing the future patella (*P*). Goldner, × 100

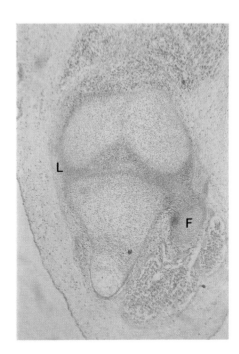

Fig. 13.2. Embryo 123N, stage 19, 7 weeks, 19 mm. Frontal section. Owing to the flexed position, both femoral condyles are cut transversely whereas tibia and fibula (*F*) are seen in a frontal plane. Formation of collateral ligaments (*L*) is suggested by dense-celled areas. The quadriceps and muscles in the anterolateral compartment of the leg are forming. The interzone begins to assume a three-layered appearance. Goldner, × 40

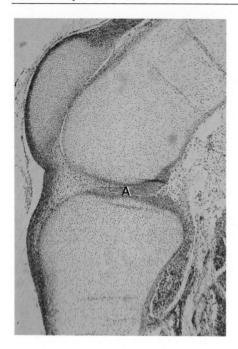

Fig. 13.3. Embryo 50N, stage 23, 8 weeks, 30 mm. Sagittal section. The quadriceps tendon, the patellar and anterior (*A*) cruciate ligaments, and the beginning of patello-femoral joint cavitation are well outlined. Loose cellular tissue occupies the future site of the infrapatellar fat pad. Whereas the shape of the proximal tibia resembles that of an adult bone, modeling of the femur has not yet started. HPS, ×40

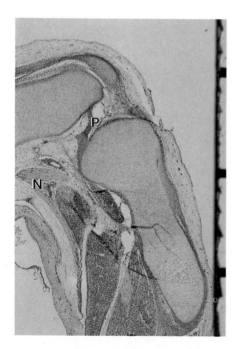

Fig. 13.4. Fetus 70N, 9½ weeks, 45 mm. Sagittal section. The extensive joint cavitation, the posterior cruciate (*P*), the sciatic nerve (*N*) just at its division, and the popliteal artery are well seen. The patellar tendon is continuous with the fibrous layer of the perichondrium. Posterior to it, loose cellular, highly vascularized tissue is seen which will become the fat pad. Goldner, ×4.3

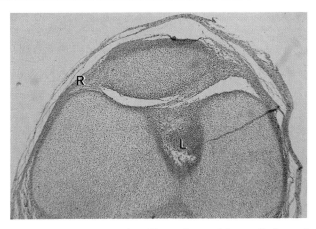

Fig. 13.5. Fetus 51N, 9½ weeks, 45 mm. Transverse section. The surfaces of the patellofemoral joint are irregular. At the level of the medial facet soft tissue strands are still present. The lateral retinaculum (*R*) is seen as a dense structure, the medial one is much less dense, but broader. The femoral insertion of the posterior cruciate ligament (*L*) is evident. Goldner, ×40

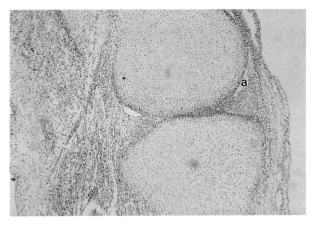

Fig. 13.6. Fetus 67, 9½ weeks, 50 mm. Sagittal section. Cavitation between femur and anterior (*a*) and posterior horn of the lateral meniscus has started. It always precedes the cavitation between meniscus and tibia. HPS, ×40

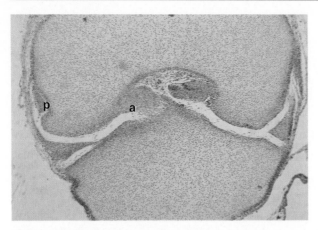

Fig. 13.7. Fetus E6, 10 weeks, 50 mm. Frontal section. The popliteal tendon (*p*) identifies the lateral condyle. The lateral meniscus, the anterior (*a*) and posterior cruciate ligaments, and the medial meniscus are distinct. Note that the lateral meniscus is not discoid. The collateral ligaments are well formed. Goldner, × 40

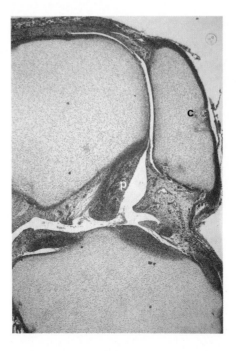

Fig. 13.8. Fetus 106N, 12 weeks, 70 mm. Sagittal section. Cartilage canals (*c*) entering the patella can be seen. The joint cavity reaches under the quadriceps. A strong posterior cruciate (*p*) ligament is well seen, as is the loose infrapatellar tissue. The patellar ligament continues into the fibrous perichondrium. Azan, × 25

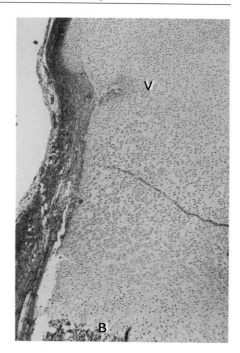

Fig. 13.9. Same fetus as in Fig. 13.8. Beginning of the formation of the anterior tibial tuberosity as seen by a vessel (*V*) invading the proximal tibial cartilage anlage in an oblique direction. Note the distance to the advancing enchondral bone formation (*B*), which is decreasing with increasing age. Compare with Figs. 13.10 a and 13.11. Azan, ×40

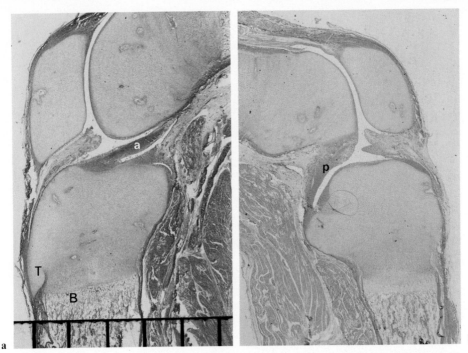

a

b

Fig. 13.10. a Fetus 40N, 14 weeks, 110 mm. Sagittal section. The anterior cruciate ligament (*a*) is surrounded by the joint space. A covering by synovial-like cells is seen. This is in contradistinction to the posterior cruciate ligament, seen in Fig. 13.10 b which seems to be extra-articular. The tibial enchondral bone formation (*B*) has almost reached the level of the anterior tibial tubercle (*T*) (see Fig. 13.11). HPS, scale in millimeters **b** Same fetus as in Fig. 13.10 a. Sagittal section. This section is slightly more medial and shows the posterior cruciate ligament (*p*). The enchondral ossification of the tibia has reached the proximal metaphysis. Goldner, same magnification as in Fig. 13.10 a

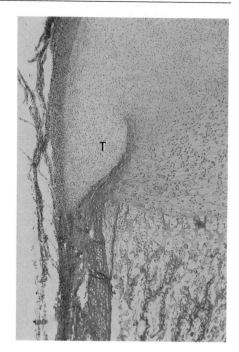

Fig. 13.11. Same fetus as in Fig. 13.10. Sagittal section. This microphotograph shows the progression of the vascular invasion of the proximal tibial epiphysis. The columns of chondrocytes indicate that the physis is now forming. The fibers of the patellar ligament do not seem to attach to the tongue-shaped cartilage, but are in continuity with the periosteum more distally. Tibial tubercle (*T*). HPS, ×40

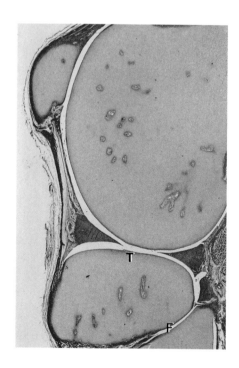

Fig. 13.12. Fetus 13N, 16 weeks, 135 mm. Sagittal section. This section goes through the lateral compartment. It shows the femoral modeling, the tibiofibular (*F*), femorotibial (*T*), and patellofemoral joints. Note also the vascularity at the peripheral attachment of the meniscus (see also Fig. 13.13). HPS, ×8

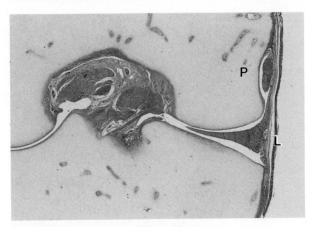

Fig. 13.13. Fetus 148N, 16 weeks, 136 mm. Frontal section. This specimen permits a closer study of the lateral collateral ligament (*L*) which is seen separated from the lateral meniscus by loose cellular tissue. Note also the popliteal tendon (*P*). This section crosses the knee in its more posterior part; this explains the elongated shape of the lateral meniscus. It should not be interpreted as representing a discoid meniscus. HPS, × 3

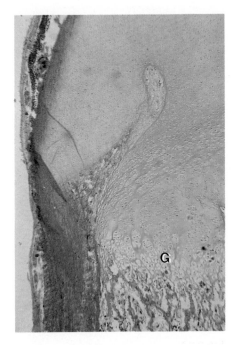

Fig. 13.14. Fetus 73N, 18 weeks, 110 mm. Sagittal section. The formation of the tibial tuberosity has progressed considerably. The proximal tibial growth plate (*G*) is now well organized. Goldner, × 40

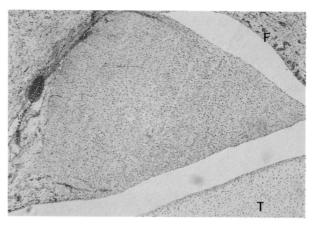

Fig. 13.15. Fetus 77, 19½ weeks, 180 mm. Sagittal section. The medial meniscus consists of dense collagenous, but not of fibrocartilaginous tissue. Note the richly vascularized part of the meniscus at its periphery. Tibia (*T*); fatpad (*F*). HPS, × 50

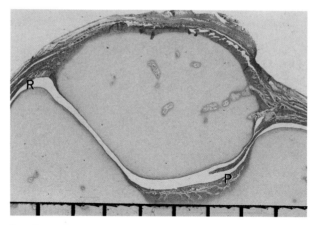

Fig. 13.16. Fetus 115N, 20 weeks, 185 mm. Transverse section. The patellofemoral joint is cut at its distal part. A plica (*P*) is seen coming in from the medial side and is in continuity with the medial retinaculum. Note the dense lateral retinaculum (*R*). HPS, scale in millimeters

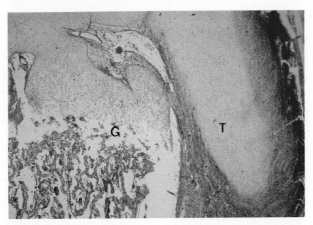

Fig. 13.17. Fetus 116N, 20½ weeks, 190 mm. Sagittal section. The proximal tibial growth plate (*G*) has grown beyond the tip of the cartilaginous tongue of the anterior tibial tubercle (*T*). HPS, ×25

References

1. Gardner E, O'Rahilly R (1968) The early development of the knee joint in staged human embryos. J Anat 102:289–299
2. Gray DJ, Gardner E (1950) Prenatal development of the human knee and superior tibiofibular joints. Am J Anat 86:235–288
3. Hughes ES, Sunderland S (1946) The tibial tuberosity and the insertion of the ligamentum patellae. Anat Rec 96:439–444
4. Ogata S, Uhthoff HK (1990) The development of synovial plicae in human knee joints. An embryologic study. Arthroscopy 6 (in press)
5. Ogden JA, Hempton RF, Southwick WO (1975) Development of the tibial tuberosity. Anat Rec 182:431–446
6. Wassilev W (1972) Elektronenmikroskopische und histochemische Untersuchungen zur Entwicklung des Kniegelenkes der Ratte. Z Anat Entwicklungsgesch 137:221–238

Chapter 14

The Development of the Ankle and Foot

T. Kawashima and H.K. Uhthoff

Admittedly the visualization of a three-dimensional structure from serial histologic sections is not easy. Moreover, changes in the position of the legs and feet during embryonal and early fetal periods render the interpretation even more difficult. Therefore, before starting the histologic description of the development, we would like to explain briefly the positional changes of leg and foot.

Lower limb buds are first seen about 4 weeks of age (stages 12 and 13; see Chap. 1) and by the end of the 6th week (stage 17) the distal portion of the limb bud develops into the digital plate. Initially digital plates are oriented in line with the long axis of the lower leg. Because of the external rotation of the whole lower limb, the plantar surface of the digital plate is facing in a cranial direction (Fig. 14.1 a, b). The digital plate becomes notched at the beginning of the 8th week (stage 20; Fig. 14.1 c, d). The notches become deeper and this gives rise to a fan-like appearance of the digital plate at stage 21. As external rotation decreases, the plantar surface faces more medially than at earlier stages.

The degree of plantar flexion (equinus) starts to decrease from stage 22 which makes more evident the inversion (varus) and adduction of the foot. At the end of the 8th postovulatory week (stage 23) the plantar surfaces face each other and the feet are in a position of equino-varus-adductus (Fig. 14.1 e, f).

The degree of equinus, varus, and adductus decreases gradually during the 10th–11th week. The feet will reach an almost neutral position at the end of the 11th postovulatory week [2]. These positions cannot be attributed to a restriction of movement as severe restrictions of movements of the fetus in the amniotic fluid usually start at about 6 months. Therefore, the positional changes of the feet during earlier periods seem to depend mostly on the preprogrammed skeletal and neuromuscular development. Moreover, joint cavities with synovial lining do not appear before 10 weeks.

From the histologic point of view, skeletal development of the lower leg and foot proceeds in a proximodistal direction. Mesenchymal condensations of the future tibia and fibula appear at stage 17, and of future 18. These mesenchymal (blastemic) tissues begin to chondrify a few days later (Fig. 14.2) and by stage 23 all elements except sesamoids are usually chondrifying (Figs. 14.5 and 14.6).

As the maturation of the cartilage progresses, a periosteal bony collar starts to form around the shaft of long bones, found around the tibia at stage 22 (Fig. 14.5), around the fibula at stage 23, and around the metatarsi at $8\frac{1}{2}-10\frac{1}{2}$ weeks (Fig. 14.10).

Vascular invasion of the cartilage anlage occurs several days after completion of the bone collar and phase 1 of enchondral bone formation starts in the tibia at 9 weeks, in the fibula at 10 weeks, and in metatarsi at 11 weeks, respectively.

Whereas proximal and middle phalanges ossify like other short tubular bones, ossification of the distal phalanges is different. In those bones, intramembranous ossification caps the cartilaginous anlage of the distal phalanx at the 11th week. Vascular invasion of the cartilage anlage is followed by enchondral bone formation which takes place between the cartilaginous anlage and membranous bone formation capping it. Enchondral bone formation advances in a proximal direction only (Figs. 14.11, 14.12, and 14.16).

The calcaneus is the first tarsal bone to ossify. Initially, periosteal bone formation starts at the lateral aspect of the calcaneus at 13–14 weeks (Figs. 14.13 and 14.14) and then the enchondral bone formation starts under it by the 15th–16th postovulatory week. Talus and other tarsal bones usually begin to ossify at 7 months or later. The first stage of joint development consists of a densely cellular zone (homogeneous interzone) which is found between chondrifying skeletal elements (Figs. 14.2, 14.3, and 14.4).

By the end of the embryonic period proper a three-layered interzone in the larger joints develops. As elsewhere it is composed of two chondrogenic dense cellular layers separated by a loose zone (destined to become a joint cavity; Fig. 14.5). Simultaneously with the appearance of the three-layered interzone the cellular condensation of the future capsule appears at the periphery (Figs. 14.4 and 14.5).

Cavitation (small isolated cavities or clefts in the loose-celled layer) of the ankle is usually seen at the 9th week and the whole joint cavity with synovial lining is present by 10–11 weeks (Fig. 14.11). Finally, a joint cavity with synovial villi begins to form at 14 weeks in most of the tarsal joints (Figs. 14.13 and 14.15).

By the end of the embryonic period all of the major muscular, neural, and vascular elements of the limb are present in a form and arrangement closely resembling those of the adult (Figs. 14.6 and 14.9). Tendons are usually differentiated by stage 20 (Fig. 14.2) and their synovial sheaths, separated from the tendon by a cavity, appear by stage 23 (Fig. 14.6).

Initially, the insertion of tendons and ligaments does not show the clear anchorage of collagen fibers into the cartilaginous anlage. Usually, tendons attach to the thick fibrous perichondral layer, i.e., collagen fibers of a tendon or ligament blend into the fibers of the perichondrium, but never anchor in the cartilage itself during the period under study (before 20 weeks; Fig. 14.16).

During the embryonic and fetal periods, cartilaginous bridges are often found between talus and calcaneus, usually at the level of the sustentaculum tali and between the 3rd metatarsus and 3rd cuneiform [3]. But it is not sure whether any of those bridges will persist and eventually form a tarsal coalition at the age of skeletal maturity or not (Fig. 14.8).

We would like to stress some other important changes in the position of the elements of ankle and foot during normal development [2]. At an early age (Figs. 14.2, 14.4, 14.5, and 14.6), and obliquity in two planes of the tibiotarsal joint is noted, responsible for equinus and inversion. With normal development both obliquities will disappear (Fig. 14.15). Also, a fan-shaped appearance of the metatarsi is noted (Fig. 14.3). The fan-shaped position of metatarsi II–V will disappear rather early, leaving for a certain period an adduction of the first metatarsus (Fig. 14.7). This will also disappear with time (Fig. 14.10).

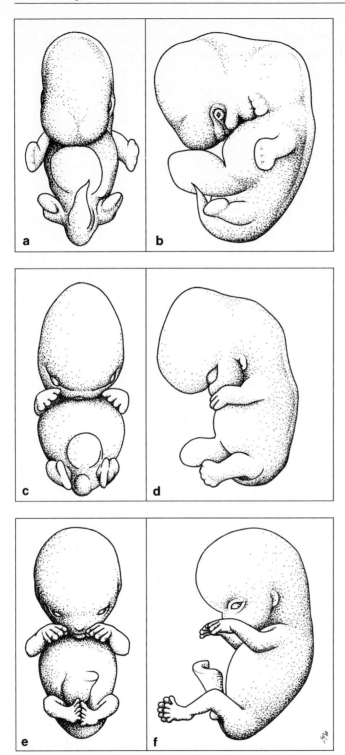

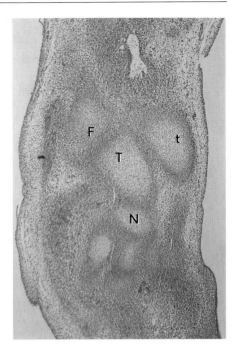

Fig. 14.2. Embryo 273, 7 weeks, stage 20, CR 20 mm. Foot in extreme equinus. Chondrification of the tarsus and metatarsus has taken place. A densely cellular zone separates each anlage from the other at sites of future joints. Note the lateral inclination of distal tibial articular surface. Fibula (*F*); tibia (*t*); talus (*T*); navicular (*N*). HPS, ×40

Fig. 14.1 a–f. a, b Stage 17. The distal portion of the lower limb buds develops into the digital plate by this stage. The plantar surface of the plate is facing in a cranial direction. Finger rays have already appeared in the hand plate. Upper and lower limb buds are externally rotated as a whole **c, d** Stage 20. The digital plate becomes notched by this stage. Distinct interdigital spaces are already seen in the hand plate. The plantar surface undergoes rotation to face more medially than at earlier stages **e, f** Stage 23. The foot soles are facing each other on the midsagittal plane with equino-varus-adductus. At this stage the big toe is usually adducted (hallux varus) at the level of the metatarsophalangeal joint. [Modified from 7]

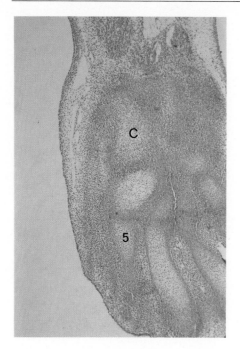

Fig. 14.3. Same embryo as in Fig. 14.2. This section is more plantar. The metatarsi are further apart than at a later stage. The cartilaginous cells are in Streeter phase 2. Calcaneum (*C*); fifth metatarsus (*5*). HPS, × 40

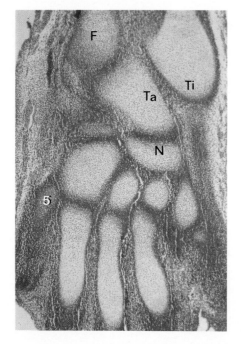

Fig. 14.4. Embryo 266, 7½ weeks, stage 21, CR 24 mm. Because of the extreme plantar flexion, normal for this stage of development, foot and leg are seen on the same plane. Serial sections of this specimen showed that almost all of the skeletal elements except the distal phalanges are in different stages of chondrification. A homogeneous cellular interzone can be seen in the ankle and other tarsal joints. At the base of the metatarsi the early formation of collateral ligaments is evident. Note the lateral inclination of the distal tibial articular surface. Fibula (*F*); talus (*Ta*); navicular (*N*); fifth metatarsus (*5*); tibia (*Ti*). HPS, × 40

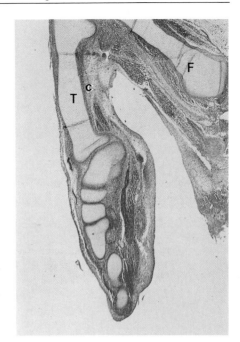

Fig. 14.5. Embryo 265, 7½ weeks, stage 22, CR 27 mm. Severe plantar flexion normal for this stage of development is well demonstrated. Note the posterior inclination of the distal tibial articular surface. The three-layered interzone at the ankle joint is easily recognized in this section, but other tarsal joints are still homogeneous. Note the posterior capsule of the tibiotalar joint. Femur (*F*); tibia (*T*); bony periosteal collar (*C*). HPS, × 20

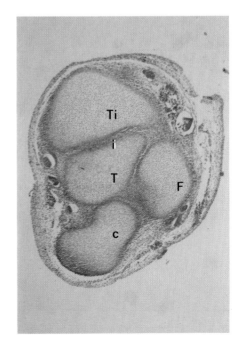

Fig. 14.6. Embryo 306, 8 weeks, stage 23, CR 30 mm. The ankle joint has developed into a three-layered interzone (*i*); note the loose tissue at the periphery of the future joint: tibia (*Ti*), talus (*T*), calcaneus (*c*), lateral malleolus (*F*). Tibialis posterior, flexor hallucis longus and flexor digitorum longus tendon (left, from the top), peroneus longus and brevis tendon (right) are observed in their tendon sheath. Note the calcaneovarus position normally seen at this stage. Toluidine blue, × 40

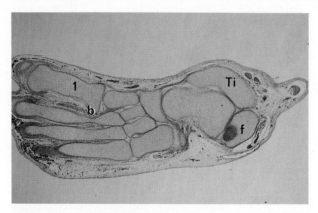

Fig. 14.7. Fetus 254, 9 weeks, CR 39 mm. A bone collar (*b*) starts to surround the second metatarsus. Note the fibrous capsule as well as vascular invasion of the loose connective tissue at the periphery of the first tarsometatarsal joint, but a joint cavity with the synovial lining has not appeared yet. The first metatarsus (1) still shows some adduction at the tarsometatarsal joint. Tibia (*Ti*); fibula (*f*). H & E, × 40

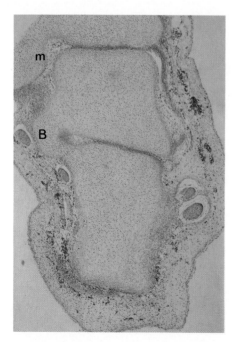

Fig. 14.8. Fetus 251, 9 weeks, CR 48 mm. Frontal section. Section at the level of the sustentaculum tali shows a cartilaginous bridge (*B*) (the bridge was not located intra-articularly, but extra-articularly between the posterior part of the sustentaculum tali and the posteromedial tubercle of the talus). Medial malleolus (*m*). HPS, × 40

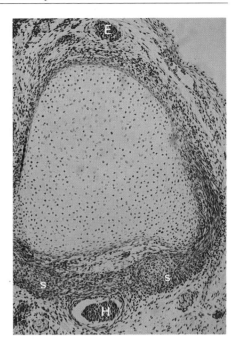

Fig. 14.9. Fetus 313, 9 weeks, CR 44 mm. This section cuts through the first metatarsal head and sesamoids (*s*) at the level of the metatarsophalangeal joint. Tibial and fibular sesamoids and intersesamoidal ligament are well demonstrated. Flexor hallucis longus (*H*) is seen in its well-developed tendon sheath at the plantar aspect. Neurovascular bundles are also seen at the plantar side superficial to the sesamoids. The extensor hallucis longus (*E*) is not surrounded by tendon sheath. Azan, × 100

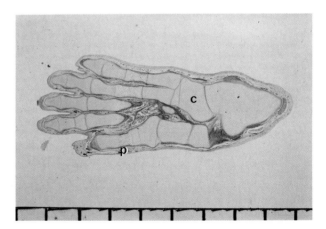

Fig. 14.10. Fetus 261, 10½ weeks, CR 60 mm. At this stage alignment of the tarsal bones resembles more or less the "adult" configuration. Although the shaft of the first metatarsus contains hypertrophic cartilage cells, vascular invasion the metatarsus has not yet occurred. The periosteal bone formation (*p*) is particularly well seen at the first metatarsus. Cuboid (*C*). Azan, scale in millimeters

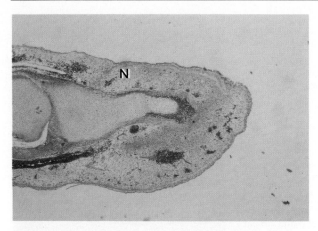

Fig. 14.11. Fetus 219, 10½ weeks, CR 62 mm. The membranous ossification has taken place at the distal part of the distal phalanx. The joint cavity is lined with synovial tissue. The dorsal surface of the toe differentiates into a primary nail field (*N*) which constitutes the earliest external change in the development of the nail [8]. The proximal border of nail field forms the proximal nail fold. Azan, × 40

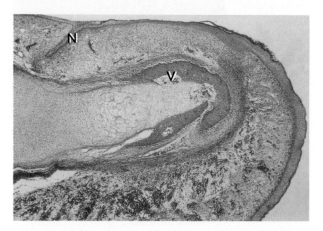

Fig. 14.12. Fetus 107, 14 weeks, CR 105 mm. Same level as section in Fig. 14.11. Vascular invasion (*V*) of the cartilage starts from the distal part covered by the intramembranous bony cap. At this age the nail plate (*N*) is recognizable. H & R, × 100

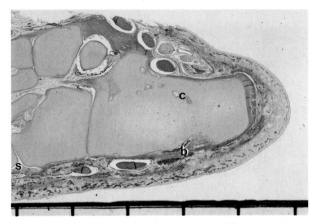

Fig. 14.13. Fetus 173, 14 weeks, CR 105 mm. Synovial villi (*s*) can be found in the midtarsal joints. This section shows cartilage canals (*c*) only in the calcaneus. Usually cartilage canals are found in all the tarsal bones after 14 weeks. Note the periosteal bone formation (*b*) at the lateral part of the calcaneus. Azan, scale in millimeters

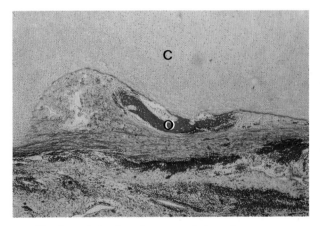

Fig. 14.14. Fetus 173, higher magnification of the lateral aspect of the calcaneus, an area identified as *b* in Fig. 14.13. Periosteal ossification (*O*) of the calcaneus (*C*) starts at the lateral aspect. Azan, ×40

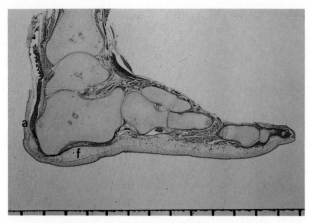

Fig. 14.15. Fetus 105, 14½ weeks, CR 112 mm. The foot is now in a neutral position. Cartilage canals are present in almost all tarsal bones by this age. A fibrous capsule and well-developed synovial villi are seen in the ankle and subtalar joints. The Achilles tendon (*a*) inserts in the thick fibrous perichondrium of the posterior part of the calcaneus; the plantar fascia (*f*) is also continuous with this thick perichondrium. H & E, scale in millimeters

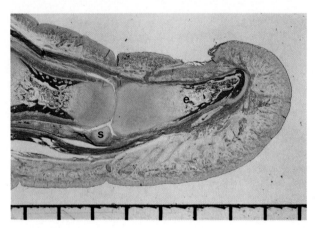

Fig. 14.16. Fetus 176, 20 weeks, CR 180 mm. Enchondral ossification (*e*) of the distal phalanx of the great toe is now predominant. Note the difference in the direction of the development of enchondral ossification between distal and proximal phalanges. A well-developed nail plate, sweat glands, tendons, joint capsule, and sesamoid are seen in this section. The sesamoid (*s*) of the interphalangeal joint is also seen, although it is not always present, even in the postnatal period (2%–4% in the adult). Note that both flexor and extensor tendons insert into the fibrous layer of the perichondrium and not into the anlage of the distal phalanx. Azan, scale in millimeters

References

1. Gardner E, Gray DJ, O'Rahilly R (1959) The prenatal development of the skeleton and joints of the human foot. J Bone Joint Surg [Am] 41:847–879
2. Kawashima T, Uhthoff HK (1990) Development of the foot in prenatal life in relation to idiopathic club foot. J Pediatr Orthop 10:232–237
3. Kawashima T, Uhthoff HK (1990) Prenatal development around the sustentaculum tali and its relation to talocalcaneal coalitions. J Pediatr Orthop 10:238–243
4. O'Rahilly R (1973) The human foot. In: Giannestras NJ (ed) Foot disorders. Medical and surgical management, 2nd edn. Lea and Febiger, Philadelphia, pp 16–23
5. O'Rahilly R, Gardner E (1975) The timing and sequence of events in the development of the limbs in the human embryo. Anat Embryol (Berl) 148:1–23
6. O'Rahilly R, Gardner E, Gray DJ (1960) The skeletal development of the foot. Clin Orthop 16:7–14
7. Streeter GL (1951) Developmental horizons in human embryos. Carnegie Institution of Washington, Washington
8. Zaias N (1963) Embryology of the human nail. Arch Dermatol 87:37–53

Chapter 15

Variations and Malformations

H.K. UHTHOFF

Vertebrae

In our collection we have observed an unsually high number of developmental abnormalities. We believe that this high incidence is related to the fact that our collection consists exclusively of spontaneously aborted embryos and fetuses. The highest number of abnormalities was seen at the level of the spine. In an earlier report [8] we described 11 malformations in 266 specimens. Since then we have found five more vertebral malformations. In all but one specimen the malformations were observed before the process of ossification had started. Our observations are limited to the level of vertebral bodies. Malformations are divided into three types: type 1, failure of formation; type 2, failure of segmentation; and type 3, mixed malformations [7]. Type 1, failure of formation, is subdivided into defect of formation and error of formation. Type 2, failure of segmentation, is similarly subdivided into defect of segmentation and error of segmentation. Defect denotes either an incomplete formation or an incomplete segmentation. Error, on the other hand, is characterized by an abnormal formation or by an abnormal direction of segmentation. Type 3 groups all malformations in which failure of formation and failure of segmentation are observed simultaneously. An example of a defect in formation is a hemivertebra (Figs. 15.1, 15.2) whereas an abnormally shaped vertebra represents an error of formation (Fig. 15.3). A defect of segmentation can be seen as an unsegmented bar (Fig. 15.4) or a block vertebra (Fig. 15.5). A V-shaped segmentation leading to an oblique intervertebral disk space (Fig. 15.6) represents an error in segmentation. Type 3, representing the simultaneous presence of failure of formation and failure of segmentation, can be seen in Figs. 15.7 and 15.8.

Vertebral segmentation is said to be induced by the notochord, and vertebral malformations are often attributed to a malformed notochord [6]. We, however, can document that the notochord in human embryos does not necessarily induce malformations [1, 2]. We observed several instances of notochordal abnormalities; in all of them the development of the vertebral bodies was normal.

In clinical practice vertebral body malformations can only be detected radiologically. It is therefore wrongly assumed that malformations occur during the stage of ossification, owing to nonvascularization of a given part of the cartilaginous vertebra [5]. Although malformations in our series were seen in all but one specimen during the stage of chondrification, they most probably develop during the earlier blastemic stage.

The occasional radiologic presence of an anterior and a posterior center of vertebral ossification should not be interpreted as a malformation. In serial sections, done in all three planes, we have consistently been able to show that a bony bridge containing a blood vessel connected two larger centers of ossification (Fig. 15.9). It seems that the bridge is often too small to be detected radiologically [9]. Interestingly, the cartilaginous maturation cranial and caudal to the bridge was delayed. This anomaly, known as coronal cleft, is a developmental variant, but not a malformation. It can be sometimes seen on lateral radiograms of fetuses and newborns as a radiolucent band between an anterior and posterior nucleus of ossification. During subsequent growth, however, this band of decreased density will disappear spontaneously. Our topographical analysis has shown that the notochord cannot be held responsible for this variant since it lies anterior to the area of delayed cartilaginous maturation [2].

Shoulder

At the level of the shoulder we did not observe any evidence of malformation. Of interest is the fact that the membranous ossification of the clavicle starts in two centers (Fig. 15.10). During normal development these centers will fuse. It has been speculated that congenital pseudarthrosis of the clavicle will develop, should this fusion fail to occur. We cannot support this view [4].

Elbow

At the elbow we observed subluxation and dislocation of the radial head (Fig. 15.11).

Wrist and Hand

No anomaly was observed at the wrist or hand.

Hip

The most impressive anomaly at the hip was the insufficient cartilaginous coverage of the femoral head by the posterior part of the acetabulum. A broad labrum partially covers the head; the capsule is not attached to its crest. Given the flexed position of the hip in utero any thrust exerted on the femur will push the head against the weak posterior wall. In one instance we could observe a posterior subluxation of the femoral head (see Chap. 12, Fig. 12.12).

Knee

We observed one fetus with bilateral rotatory subluxation at the knee (Fig. 15.12). The quadriceps muscle showed less muscle fibers, but more fibrous tissue than normal. The suprapatellar pouch was obliterated. As in reported cases of congenital dislocation in newborns, the knees were in slight hyperextension. The presence of a medial plica was seen in three instances (Fig. 15.13). This plica was always attached to the fat pad (Ogata and Uhthoff, personal observation).

Foot

Malformations of the foot were mostly limited to a bridge at the sustentaculum tali [3]. Interestingly, this bridge was seen more often at the posterior, extra-articular part (Chap. 14, Fig. 14.8) than at the anterior, intra-articular part. We also observed absence of joint formation between the third cuneiform and the base of the third metatarsus. This is often referred to as fusion, although it is rather doubtful whether a fusion between two cartilaginous anlagen has actually occurred or whether they have developed as one anlage from the beginning.

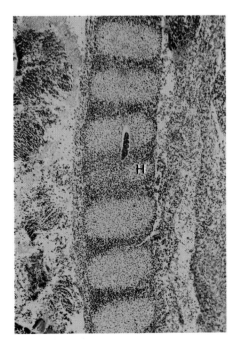

Fig. 15.1. Embryo 81, 6 weeks, CR length 13 mm, stage 17. Sagittal section. Defect of formation of D-10 leading to a hemivertebra (*H*). The anterior part is absent. HPS, × 40. [From 8]

Footnote: Figures 15.1, 15.2, 15.3, 15.4 and 15.7 are reproduced with the kind permission of Acta Orthop. Scand. Tanaka, T. and Uhthoff HK. 1981. 52:413–425

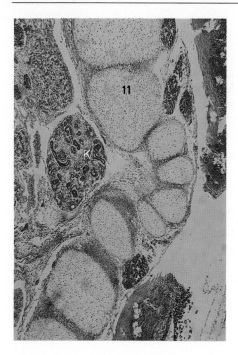

Fig. 15.2. Embryo 78, 7 weeks, CR length 20 mm, stage 20. Sagittal section. Defect of formation of six vertebrae at the thoracolumbar level. In four the anterior half is missing. The kidney (*K*) is situated anterior to the hemivertebrae, most probably representing a horseshoe kidney. The vertebrae cranial and caudal to the hemivertebrae are also malformed. D-11 (11). HPS, ×40. [From 8]

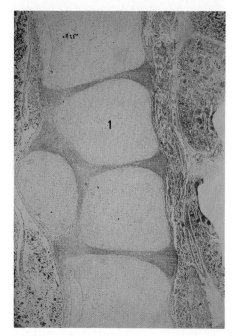

Fig. 15.3. Fetus 64, 9½ weeks, CR length 45 mm. Sagittal section. Error of formation of L-1 and L-2. The posterior part of L-1 (1) is missing whereas the posterior part of L-2 seems detached and protrudes into the spinal canal. Serial sections showed that this part was attached to L-2 by a pedicle. This seems to represent the earliest case of spinal stenosis ever reported. HPS, ×20. [From 8]

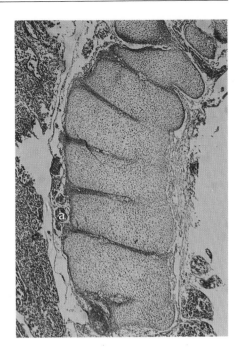

Fig. 15.4. Embryo 77, 7 weeks, CR length 20 mm, stage 20. Defect in segmentation of the upper thoracic spine leading to a unilateral unsegmented bar. Intersegmental arteries (*a*) are only seen on the side of segmentation whereas venous channels can be recognized on the side of the bar. HPS, × 40. [From 8]

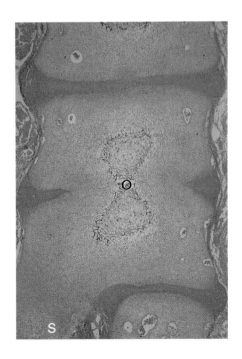

Fig. 15.5. Fetus 89N, 11½ weeks, CR length 70 mm. Frontal section. Defect of segmentation between L-4 and L-5 leading to a block vertebra. Both ossific nuclei (O) are in intimate contact. Note also that one side of L-5 is in continuity with S-1 (S). Goldner, × 15

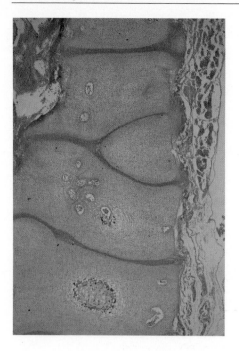

Fig. 15.6. Same fetus as in Fig. 15.5. Error of segmentation at the level of D-3 and D-4. Serial sections showed that this V-shaped segmentation reached through the entire width in the posterior part. Goldner, × 15

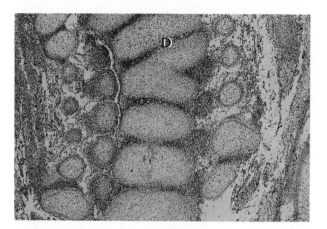

Fig. 15.7. Embryo 83, 5½ weeks, CR length 10 mm, stage 16. Frontal section. Mixed malformation of the thoracic spine. A hemivertebra forms one cartilaginous anlage with a normal vertebra. Above this level the intervertebral disk space is oriented in an oblique direction (*D*). HPS, × 40. [From 8]

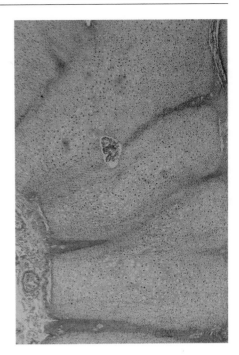

Fig. 15.8. Embryo 73, 7 weeks, CR length 19 mm, stage 19. Frontal section. A mixed malformation of the cervical spine which can be called a Klippel-Feil syndrome. No malformation of either scapula was seen. Goldner, × 40

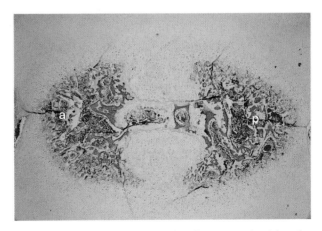

Fig. 15.9. Fetus 5, 20½ weeks, 190 mm. Sagittal section. Occasionally an anterior (*a*) and a posterior (*p*) center of ossification are present in one vertebra. They are always connected by a bony bridge containing blood vessels. Goldner, × 20

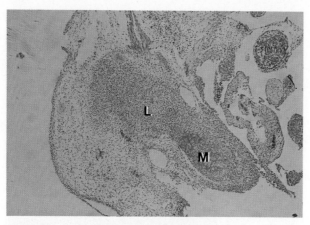

Fig. 15.10. Embryo 273, 7 weeks, CR length 20 mm, stage 20. Frontal section. The midpart of the clavicle is formed through direct bone formation. This process starts in two separate ossification centres which will fuse. The medial ossification center (*M*) is always bigger than the lateral one (*L*). Azan, × 40

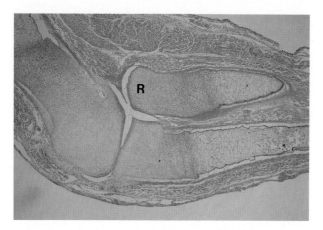

Fig. 15.11. Fetus 71N, 19½ weeks, CR length 180 mm. Sagittal section. An anterior dislocation of the radial head (*R*) is clearly seen. Goldner, × 5.5

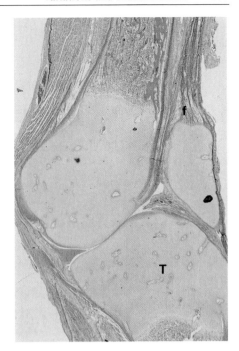

Fig. 15.12. Fetus 71N, 19½ weeks, CR length 180 mm. Sagittal section. In this fetus a rotatory subluxation of both knees was seen. The knee is in hyperextension, and the tibia (*T*) displaced forward. Although barely seen in this photograph the quadriceps muscle contained an unusual amount of fibrous elements (*f*) Goldner, × 5.5

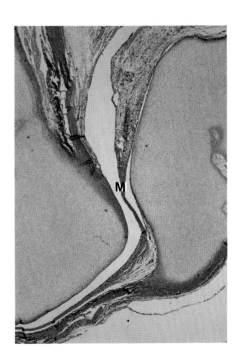

Fig. 15.13. Fetus 72, 8 weeks, CR length 27 mm, stage 23. Sagittal section. Medial plica (*M*) between patella and medial femoral condyle. Goldner, × 20

References

1. Goto S, Uhthoff HK (1985) Notochord action on spinal development. Acta Orthop Scand 57:85–90
2. Goto S, Uhthoff HK (1986) Notochord and spinal malformations. Acta Orthop Scand 57:149–153
3. Kawashima T, Uhthoff HK (1990) Prenatal development around the sustentaculum tali and its relation to talocalcaneal coalitions. J Pediatr Orthop 10:238–243
4. Ogata S, Uhthoff HK (1989) The early development and ossification of the clavicle. Acta Orthop Scand (submitted for publication)
5. Schmorl G, Junghanns H (1971) The human spine in health and disease, 2nd edn. Grune and Stratton, New York
6. Shapiro F, Eyre D (1981) Congenital scoliosis. A histopathologic study. Spine 6:107–117
7. Tanaka T, Uhthoff HK (1981) Significance of resegmentation in the pathogenesis of vertebral body malformation. Acta Orthop Scand 52:331–338
8. Tanaka T, Uhthoff HK (1981) The pathogenesis of congenital vertebral malformations. Acta Orthop Scand 52:413–425
9. Tanaka T, Uhthoff HK (1983) Coronal cleft of vertebrae, a variant of normal enchondral ossification. Acta Orthop Scand 54:389–395

Subject Index

J. Wolff

The Law of Bone Remodelling

Translated from the German by P. Maquet, R. Furlong

1986. XII, 126 pp. 95 figs. Hardcover DM 198,–
ISBN 3-540-16281-X

Julius Wolff's book is mentioned in about every work dealing with bone structure. His name is familiar to every anatomist and orthopaedic surgeon. Very few, however, have actually read his book, written in the flowery scientific German of the turn of the century. A translation into English was thus needed to make the work available to a larger audience in the problems of bone structure.

B. N. Tillmann, University of Kiel

Slides in Human Arthrology

1985. 207 slides (legends in English). 36 pp. DM 578,–
ISBN 3-8070-0350-9

This ring binder contains 207 colored slides of diarthroses of the extremities and trunk. First, figures of the bony parts of the joints and the cartilage-covered articular surfaces are presented, followed by the capsular ligaments, intra-articular structures, and prepared sections of joints. In addition to normal findings, frequently occurring structural anomalies are also shown. Each of the three divisions also contains figures of typical degenerative changes in the joints. The slides are accompanied by a booklet with explanatory text identifying visible structures in each slide. Simplified labeled drawings aid identification where deemed necessary.

Springer-Verlag Berlin
Heidelberg New York London
Paris Tokyo Hong Kong

Preisänderungen vorbehalten

Springer

J. Guyot, University of Besançon, France

Atlas of Human Limb Joints

2nd rev. ed. 1990. X, 258 pp. 116 figs. Hardcover DM 312,-
ISBN 3-540-51709-X

In this work, the author provides the most complete description of human limb joints available today. His presentation is divided into two parts. The first part contains a summary of the functional anatomy of each of the joints.
The second part is devoted to the pictorial illustration of the joints, consisting of photographs, drawings and diagrams of meticulously prepared dissections of the ligamentous structures surrounding the joints as well as the joints themselves.

From the reviews of the first edition: "...of great importance and interest for anatomists, surgeons, specialists in sports medicine and physiotherapists, and departmental libraries must include this book. The quality of dissections, photographs and artistic diagrams must be seen to be believed. The book is highly recommended and will be of great delight to those concerned with the function and surgery of joints."

W. Braune, O. Fischer

Determination of the Moments of Inertia of the Human Body and its Limbs

Translated from: Abhandlungen der mathematisch-physischen Classe der Königlichen Sächsischen Gesellschaft der Wissenschaften, Vol. XVIII, no. VIII, pp 407-492

Translated from the German by P. Maquet, R. Furlong

1988. VIII, 84 pp. 12 figs. 15 tabs. Hardcover DM 85,-
ISBN 3-540-18813-4

This is another classic contribution by Braune and Fischer to the field of biomechanics, translated here for the first time from the original German edition of 1892.
The pendulum method was employed to ascertain accurately the moments and radii of inertia of the human body and its different parts about all axes – transverse, oblique or longitudinal. This elegant method is described in detail, together with the results. Relations were found between the centres of inertia on the one hand, and the lengths and diameters of the body segments on the other.
Such work is the basis for solving the mechanical problems related to any movement of the human body: thus, the original results presented here continue to be of immense value to current research and practice.

Springer-Verlag Berlin
Heidelberg New York London
Paris Tokyo Hong Kong